Global Governance of Hazardous Chemicals

Politics, Science, and the Environment
Peter M. Haas and Sheila Jasanoff, editors

Global Governance of Hazardous Chemicals
Challenges of Multilevel Management

Henrik Selin

The MIT Press
Cambridge, Massachusetts
London, England

For information about special quantity discounts, please e-mail special_sales@ mitpress.mit.edu

This book was set in Sabon by the MIT Press. Printed and bound in the United States of America.

Library of Congress Cataloging-in-Publication Data

Selin, Henrik, 1971–
Global governance of haxzardous chemicals : challanges of multilevel management / Henrik Selin.
   p.   cm.—(Politics, science, and the environment)
Includes bibliographical references and index.
ISBN 978-0-262-01395-6 (hardcover : alk. paper)   ISBN 978-0-262-51390-6 (pbk. : alk. paper)
1. Hazardous substances—Management—International cooperation. 2. Hazardous substances—Environmental aspects—Management—International cooperation. I. Title.
T55.3.H3S44 2010
363.17—dc22
                                                                    2009029577

10   9   8   7   6   5   4   3   2   1

Contents

Acknowledgments

This book is the result of a long-standing professional and personal interest in sustainable development. It brings together in a single volume much of my research on global governance, institutions, and chemicals politics and policy making that I have been engaged in since the mid-1990s.

I have been privileged to work in several academic environments that have helped me better understand interconnected natural science and social science aspects of chemicals management issues. These include the Department of Water and Environmental Studies at Linköping University, the Environmental Policy and Planning Group at the Massachusetts Institute of Technology, the Belfer Center for Science and International Affairs and the Center for International Development at Harvard University, and the Department of International Relations at Boston University. In all of these places, I have been fortunate to meet people who have acted as great advisors, mentors, and inspirations.

Over the years, my research on global, regional, and domestic chemicals politics and management has been funded by grants from different funding agencies and organizations. In particular, I thank the Swedish Foundation for Strategic Environmental Research, the Knut and Alice Wallenberg Foundation, and the European Commission for enabling me to travel and conduct empirical research at international political and scientific meetings and meet with a wide range of policy makers and policy advocates. In addition, I am grateful to the Boston University Frederick S. Pardee Center for the Study of the Longer-Range Future for the support that helped me to conclude writing this book.

A long list of individuals has contributed to this book in a multitude of important ways. I thank the many dedicated people in international organizations, national governments, and nongovernmental organizations who have taken time out of their busy schedules to share with me their knowledge of chemicals issues. I thank Clay Morgan, Laura Callen, and Sandra Minkkinen at the MIT Press for their guidance and professionalism throughout the publishing process. I am greatly indebted to several people who provided much valuable advice and offered many insightful comments on earlier versions, including Beth DeSombre, Stacy VanDeveer, Pia Kohler, and Adil Najam, as well as three anonymous reviewers.

Finally, I recognize the encouragement provided by my family, especially Noelle. This book is dedicated to her for all her love and support, for going with me to all those Bruce Springsteen shows, and for sharing her passion for the Red Sox.

List of Acronyms

ACC	American Chemistry Council
AMAP	Arctic Monitoring and Assessment Programme
CEFIC	European Chemical Industry Council
CLRTAP	Convention on Long-Range Transboundary Air Pollution
CMA	Chemical Manufacturers Association
COP	Conference of the Parties
CSD	Commission on Sustainable Development
DDT	Dichlorodiphenyl trichloroethane
DNOC	Dinitro-ortho-cresol
EMEP	Co-operative Programme for Monitoring and Evaluation of the Long-range Transmission of Air Pollutants in Europe
EU	European Union
FAO	Food and Agriculture Organization
GEF	Global Environment Facility
HCB	Hexachlorobenzene
HCBD	Hexachlorobutadiene
HCH	Hexachlorocyclohexane
ICC	Inuit Circumpolar Council
ICCA	International Council of Chemical Associations
IFCS	Intergovernmental Forum on Chemical Safety
IGO	Intergovernmental organization
ILO	International Labour Organization
INAC	Indian and Northern Affairs Canada

IOMC	Inter-Organization Programme for the Sound Management of Chemicals
IPCS	International Programme on Chemical Safety
IRPTC	International Register of Potentially Toxic Chemicals
MARPOL	International Convention for the Prevention of Pollution from Ships
NGO	Nongovernmental organization
Octa-BDE	Octabromodiphenyl ether
OECD	Organisation for Economic Cooperation and Development
PAH	Polycyclic aromatic hydrocarbon
PBB	Polybrominated biphenyl
PCB	Polychlorinated biphenyl
PCN	Polychlorinated naphthalene
PCP	Pentachlorophenol
PCT	Polychlorinated terphenyl
PeCB	Pentachlorobenzene
Penta-BDE	Pentabromodiphenyl ether
PFOS	Perfluorooctanesulfonate
PIC	Prior informed consent
POP	Persistent organic pollutant
REACH	Regulation on Registration, Evaluation, Authorisation and Restriction of Chemicals
RoHS	Directive on the Restriction of the Use of Certain Hazardous Substances in Electrical and Electronic Equipment
SAICM	Strategic Approach to International Chemicals Management
SCCP	Short-chain chlorinated paraffin
SEPA	Swedish Environmental Protection Agency
UNCED	United Nations Conference on Environment and Development
UNDP	United Nations Development Programme
UNECE	United Nations Economic Commission for Europe
UNEP	United Nations Environment Programme

UNIDO	United Nations Industrial Development Organization
UNITAR	United Nations Institute for Training and Research
WEEE	Directive on Waste Electrical and Electronic Equipment
WHO	World Health Organization
WSSD	World Summit on Sustainable Development
WTO	World Trade Organization

1

Global Governance and the Chemicals Regime

Hazardous chemicals pose significant environmental and human health risks. A few examples demonstrate the seriousness of the situation. The average nine-year-old male beluga whale in the St. Lawrence estuary has high enough concentrations of polychlorinated biphenyls (PCBs) to be treated as a hazardous waste under Canadian legislation (Béland et al. 1993). Since a chemical park with twenty-five companies opened in Wuli Village in eastern China in 1992, it has become one of possibly several hundred Chinese "cancer villages," with a rapid surge in cancer-related illnesses and deaths (Tremblay 2007).[1] One study found that approximately 8,000 patients were admitted to a single hospital in the Indian state of Andhra Pradesh with severe pesticide poisoning between 1997 and 2002. More than 20 percent (over 1,800 people) of these patients died as a result of this exposure. Recent estimates of global pesticide poisoning put annual fatality figures close to 300,000, with 99 percent of deaths occurring in developing countries (Srinivas Rao et al. 2005).

These cases from different geographical regions, confirmed by numerous scientific and policy studies, illustrate that countries around the world face considerable difficulties establishing effective policies and administrative structures for managing hazardous chemicals. National policy makers and regulators are tasked with developing and implementing chemicals policy in the face of a host of scientific uncertainties. Many important decisions must be made based on limited scientific assessment information in situations of competing political and economic interests. Recognizing the importance of improved management of hazardous chemicals, governments at the 2002 World Summit on Sustainable Development (WSSD) in Johannesburg, South Africa, adopted the goal that chemicals should be "used and produced in ways that lead to the mini-

mization of significant adverse effects on human health and the environment" by 2020 (WSSD 2002, para. 23). Achieving this goal is a critical
but difficult governance challenge.

Global cooperation is necessary to address the full range of environmental and human health risks stemming from hazardous chemicals, as
many important issues fall within the realm of international law. For
example, the transboundary transport of persistent organic pollutants
(POPs), a specific category of particularly harmful chemicals, results in
widespread environmental dispersal far from original emission sources.
Reports by the Arctic Monitoring and Assessment Programme (AMAP)
(2002, 2004, 2009) state that many Arctic species, particularly those at
the upper end of long marine food webs, carry high levels of POPs and
that most of the POPs found in the Arctic environment come from distant
sources as emissions travel long distances. Furthermore, AMAP assessments have concluded that subtle health effects are occurring in Arctic
human populations as a result of chemical contamination of food sources.
Reports express the greatest concern for fetal and neonatal development
risks (AMAP 2003, 2009).

Many people exposed to chemical risks work in agriculture, manufacturing, or waste recovery, including the rapidly growing business in
handling electronics wastes (e-wastes). All of these sectors have a strong
connection between environmental and human health risks and the international trade in chemicals, goods, and wastes. As a result, environment-
and trade-related measures on hazardous chemicals—sometimes falling
under the legal jurisdiction of separate agreements and organizations—
need to be considered in tandem to design appropriate policy and management measures. Multilateral cooperation may also generate increased
awareness and diffuse knowledge about the severity and scope of the
chemicals problem. Many people who are exposed to hazardous chemicals are unaware of the risks and not trained to take even the most basic
risk-reduction measures. International collaboration can be important for
promoting education and supporting capacity building.

Countries that recognize domestic problems with hazardous chemicals
and want to take action often have difficulties mustering the technical,
financial, and human resources needed to initiate more effective risk-
reduction measures. This is particularly true for many developing countries.
For example, the government of Tanzania reported in a 2005 national

assessment that POP waste management facilities, including those for storage, transportation, and disposal, were basically nonexistent. Staff working with equipment possibly containing PCBs did not use any kind of protective gear. Tanzanian government officials also noted that spillage of transformer oil likely to contain PCBs was frequent and that waste transformer oil was habitually kept in open areas, or was burned or discharged "haphazardly into the environment" (Tanzania 2005, 92). This situation, not unique to Tanzania, reflects the types of problems handling hazardous chemicals that many developing countries face.

Given the management difficulties plaguing many countries, international legal, political, scientific, and technical activities to mitigate environmental and human health problems of hazardous chemicals can ideally function as catalysts for the diffusion of knowledge and resources for more effective regional and national management. Improved international and domestic chemicals management may help to reduce negative environmental and human health effects stemming from the use and mishandling of hazardous chemicals, for example. Cooperative efforts can also help prevent additional environmental dispersal of dangerous chemicals through the dissemination of alternative techniques and chemical substitutes. These and other management improvements are badly needed. Data from all over the world demonstrate that we are a long way from chemical safety, despite the fact that we are rapidly approaching the 2020 target for the safe production and use of chemicals adopted at the WSSD.

The Regime for Chemicals Management

The chemicals regime, designed to mitigate environmental and human health problems, has received little scholarly attention despite its importance.[2] It is, however, one of the oldest environmental regimes, having been in continuous development since the 1960s (and scattered international actions on hazardous substances have been taken for over a century). Countries in close collaboration with a multitude of organizations have expanded the chemicals regime to include regulations on the full life cycle of production, use, trade, and disposal of a limited number of industrial chemicals and pesticides, as well as emission controls on byproducts of production and combustion processes. The chemicals regime also contains provisions and management programs designed to assess and regulate additional chemicals, increase and harmonize information about commercial and discarded chemicals traded across countries, and augment regional and local management capacities. In addition, regime participants have established supportive organizational structures to aid implementation and regime development.

The chemicals regime is structurally different from many other major regimes that follow the "convention-cum-protocol" approach (Susskind 1994). In those cases, the creation of a framework convention outlining general policy goals is followed by the development of more detailed policies codified in protocols. For example, the ozone regime was formalized through the Vienna Convention for the Protection of the Ozone Layer and later expanded by the adoption of the Montreal Protocol on Substances That Deplete the Ozone Layer and subsequent amendments.[3] The climate change regime is developed under the United Nations Framework Convention on Climate Change. Current political efforts focus on launching a successful follow-up agreement to the Kyoto Protocol, which will expire in 2012. The biodiversity regime is structured around the Convention on Biological Diversity and the Cartagena Protocol on Biosafety. In these and other similar cases, the framework convention creates a central focal point for subsequent policymaking and management efforts.

In contrast, legal and political efforts to address hazardous chemicals did not begin with a framework convention. Instead, countries have created a set of free-standing treaties that are nonhierarchical in the sense that no one treaty is supreme over the others under international law. The chemicals regime encompasses four main multilateral agreements addressing overlapping life cycle issues: the 2001 Stockholm Convention on Persistent Organic Pollutants, the 1998 Rotterdam Convention on the Prior Informed Consent Procedure for Certain Hazardous Chemicals and Pesticides in International Trade, the 1998 Protocol on Persistent Organic Pollutants to the Convention on Long-Range Transboundary Air Pollution (CLRTAP), and the 1989 Basel Convention on the Control of Transboundary Movements of Hazardous Wastes and Their Disposal. Although these treaties are formally independent, they are legally, politically, and practically connected. In addition, many other regional agreements address hazardous chemicals in multiple ways.

While there were discussions in international political forums in the 1990s about creating a framework convention to bring together existing chemicals treaties, most countries rejected this idea. Opponents argued that it would be too complex, costly, and time-consuming to negotiate a framework convention. Critics also believed that it would be a backward way of approaching governance. A framework convention was something to start with, they held, not create halfway through a long process of establishing legal commitments (Krueger and H. Selin 2002). Governments in 2006, however, adopted the Strategic Approach to International Chemicals Management (SAICM). Although SAICM is not a legally binding agreement, it shares many traits with a framework convention. It is designed as an umbrella mechanism to guide different management efforts as it outlines a plan of action toward fulfilling the 2020 goal formulated at the WSSD. To this end, it prioritizes several key issues, including increasing information and awareness of hazardous chemicals, and enhancing domestic enforcement and management capabilities.

Despite the fact that the leading legal, political, and management responses to hazardous chemicals are formally independent, the cognitive and practical reasons to regard them as part of a regime are compelling. Cognitive reasons stem from the fact that leading states, intergovernmental organizations (IGOs), and nongovernmental organizations (NGOs) perceive the major chemicals issues to be closely connected. They act and formulate policy responses and management efforts based on these conceptual linkages. This is illustrated by political and administrative decisions under separate treaties to connect related activities and by the creation of SAICM. In addition, states, IGOs, and NGOs realize that many policy and management outcomes are practically linked with actions and decisions across forums. That is, policy making and management under one instrument shape debates and actions in other policy forums. In this respect, the expansion of the chemicals regime reflects a growth in the number and scope of agreements across a range of environmental domains.

Scholars have called attention to "treaty congestion" issues (Brown Weiss 1993, 679). Growing institutional density creates a need to explore and analyze characteristics and implications of institutional linkages, as most early regime analysis focused on empirical cases consisting of a single international agreement administered by a discrete organization. The increasing number and scope of institutions create a growth in governance linkages both within and across policy arenas. Governance linkages exist when principles, norms, rules, and decisions in one forum affect activities and outcomes in another (H. Selin and VanDeveer 2003). For example, efforts to phase out the use of ozone-depleting substances overlap with action on climate change mitigation, as some chemicals that are addressed under the ozone regime are also greenhouse gases. Similarly, policy making on desertification, deforestation, and biodiversity in a host of forums intersect with one another, as well as with activities on the use of carbon sinks under the climate change regime.

In addition, states, IGOs, and NGOs interacting within and across policy forums create important actor linkages (H. Selin and VanDeveer 2003). Many of the same actors collaborate under separate environmental treaties. For example, all major participants in the ozone regime also work together under the climate change regime. Similarly, there are great overlaps in the parties, observers, and other stakeholders who are interacting under different chemicals treaties, programs, and management efforts. This includes collaboration between treaty secretariats on shared issues and administrative tasks. Importantly, all these regime participants are engaged in linkage politics to varying degrees. That is, individual or sets of regime participants at times attempt to strategically use governance and actor linkages to forward their interests and positions across policy forums. This may include trying to secure the adoption of particular policies that they favor, as well as acting to block specific policy developments that they oppose.

Many governance and actor linkages have significant implications for multilevel governance and collective problem solving. Issues of multilevel governance—involving actors operating across horizontal and vertical levels of social organization and jurisdictional authority—have become increasingly important across issue areas as interrelated governance efforts are being developed simultaneously in multiple forums ranging across global, regional, national, and local scales. Horizontal linkages operate between instruments and programs at similar levels of social organization. For example, there can be many governance and actor linkages between two or more global treaties, as seen under the chemicals regime. In addition, vertical linkages exist between instruments and management activities at different levels of social organization. For example, ample governance and actor linkages exist among global, regional, national, and local regulations and management efforts on hazardous chemicals.

To be effective, a regime must ultimately achieve its governance objective (Faure and Lefevere 2005, Underdal 2008). Although the stated policy goals of different chemicals treaties and programs are not identically worded, it is clear that successful environmental and human health protection is a core objective of the chemicals regime. The adoption of several policy statements and the WSSD 2020 goal of ensuring that chemicals are used and produced in ways that lead to the minimization of significant adverse effects on human health and the environment reinforce this regime objective. SAICM was also established to work toward this policy goal. The realization of this regime objective is fundamentally dependent on the creation of comprehensive multilevel governance structures. That is, the effectiveness of the chemicals regime depends on the ability of regime participants to develop and implement suitable policies and management structures within and across global, regional, national, and local governance scales.

As governance and actor linkages within and across governance scales create new needs for collaborative problem solving, policy makers are exploring ways to harness regulatory synergies (Chambers 2008). Many implications of institutional linkages, however, are not well understood. The chemicals regime offers a fitting and largely overlooked case to examine linkage issues relating to policy making and management also critical to many other issue areas and governance efforts. States, IGOs, and NGOs collaborating under the chemicals regime have been dealing for several decades with many of the linkage issues that other issue areas have come to grapple with only more recently. Such linkages may have both positive and negative effects on governance efforts. As a result, studying policy issues and management experiences under the chemicals regime can lead to analytical and policy-relevant insights into the characteristics and effects of growing institutional density in global governance.

Aim of the Book

The aim of this book is twofold. First, it empirically investigates and analytically examines the development, implementation, and future of the chemicals regime as a critical but understudied area of global governance. Second, drawing on this analysis, it highlights and explores issues of policy expansions, institutional linkages, and the design of effective multilevel governance. The book thus examines issues of how policy developments, driven by coalitions of regime participants, create linkages between policy instruments addressing overlapping issues. It also considers how institutional linkages shape the interests and behavior of regime participants, as well as affect policy making and management efforts across governance scales. Specifically, the book addresses three interrelated research questions, focusing on analytical themes of coalitions, diffusion, and effectiveness. International relations scholars and practitioners in multiple issue areas are well advised to think more deeply about these themes and related issues.

The first question focuses on building actor coalitions for policy change: *How do coalitions of regime participants form in support of policy expansions, and how are their interests and actions affected by institutional linkages?* International chemicals policy has been significantly expanded by a large number of states, IGOs, and NGOs since the 1960s. The interests and actions of competing coalitions of regime participants, frequently engaging in linkage politics, have driven many of these policy expansions. However, the interests and compositions of coalitions change over time, and different coalitions are established for different policy issues. In addition, cognitive and practical linkages across multilateral forums and management efforts shape coalition politics and policy outcomes. Building on this general acknowledgment, the book analyzes in more detail ways in which institutional linkages affect the interests of regime participants and decision making across policy forums.

Much early literature focused on identifying and cataloging different types of linkages between institutions, but it often overlooked the role of regime participants in linkage politics (H. Selin and VanDeveer 2003). There is a need for empirical studies that explore the actions of regime participants in creating and developing institutional linkages. Going beyond mere descriptions of institutional linkages, this book focuses on the role of coalitions of regime participants as leaders and carriers of ideas, knowledge, and policy proposals across policy venues, thereby establishing and shaping many institutional linkages. In addition, it examines how institutional linkages influence interests and actions by regime participants, another area of linkage politics that much of the literature has ignored. To this end, this book explores the role of coalitions of regime participants in linkage politics and global governance, which are also issues of interest in many other areas where policy expansions are taking place across policy venues.

The second question focuses on the establishment and development of regime components: *How do regime participants diffuse regime components across policy venues, and how are policy diffusion and expansion efforts shaped by institutional linkages?* The many states, IGOs, and NGOs collaborating on chemicals issues have established a host of principles, norms, rules, and decision-making procedures to guide individual and collective behavior. Many regime rules have been strengthened through policy expansions across related instruments. At the same time, there have been (and continue to be) notable political differences among different coalitions of regime participants about major policy issues, including controls on specific chemicals, mechanisms for monitoring and compliance, and the organization and funding of capacity building. Whereas much regime analysis focuses on policy making taking place exclusively within the confines of a particular forum, a study of the chemicals regime must take into consideration the influence of institutional linkages.

To this end, this book examines how institutional linkages influence policy-making processes driven by coalitions of participants under the chemicals regime. It focuses on the emergence and diffusion of components that have been critical in the development and implementation of different parts of the chemicals regime, such as the movement from voluntary to legally binding approaches, the role of the precautionary principle and scientific assessments in decision making and regulation, the application of common but differentiated responsibilities, the strengthening of collective rules and commitments, the design of bodies for assessments and decision making, funding issues, and the design of capacity-building structures and programs. These issues are also of analytical and practical importance in many policy areas beyond the chemicals regime, and this study illuminates how the diffusion of regime components may influence governance efforts more generally.

The third question focuses on issues of governance and regime operation: *How do institutional linkages influence the effectiveness and design of multilevel governance efforts?* At least since the United Nations Conference of the Human Environment and the establishment of the United

Nations Environment Programme (UNEP) in 1972, public officials have attempted to integrate policy measures that address different aspects of the same environmental issue (Chambers 2008). The growth in institutional linkages operating both horizontally and vertically also increases the importance of issues of institutional fit and design (Young 2002, 2008a): an effective regime needs to be well matched to the specific environmental characteristics of the issues that it seeks to address, as well as flexible enough to be able to respond to changes in physical and political conditions. Issues of linkages, fit, and design are of growing importance in many issue areas, and this book explores how horizontal and vertical linkages influence efforts to design successful governance structures between and across different levels of social organization.

On multilevel governance in the area of chemicals management, analysts and policy makers have noted the importance of closer horizontal and vertical coordination across the many instruments and organizations addressing hazardous chemicals. In fact, the UNEP Governing Council in the 1990s identified the chemicals area as a particularly suitable pilot project for exploring opportunities for clustering agreements and enhancing regulatory and management synergies between and across governance levels. This is also a key objective of SAICM. However, efforts to improve multilevel governance and strengthen regional and local management capabilities are plagued by practical difficulties and political disagreements. Many multilevel governance consequences of institutional linkages also remain understudied. Recognizing the need to further study implications of institutional linkages, this book explores issues critical to the design and implementation of effective multilevel governance efforts, empirically focusing on the management of hazardous chemicals.

Scholarly Contribution and Main Arguments

This book connects with several analytical and policy debates. First, it contributes to a small but growing literature on chemicals management. Some of these studies address the political economy of the trade in hazardous chemicals and wastes and the development of international and national policy responses (Alston 1978; Harland 1985; Boardman 1986; Forester and Skinner 1987; Paarlberg 1993; Kempel 1993; Kummer 1995; Victor 1998; Pallemaerts 1988, 2003; Asante-Duah and Nagy

1998; Krueger 1999; O'Neill 2000; Clapp 2001; Sonak, Sonak, and Giriyan 2008; Dreher and Pulver 2008; Yang 2008). Other studies examine global and regional efforts to mitigate the transboundary transport of hazardous chemical emissions, typically focusing on one or two specific treaties (H. Selin 2003, Downie 2003, H. Selin and Eckley 2003, H. Selin and VanDeveer 2004). Another area of literature addresses justice and equity issues in the context of the international trade in hazardous substances and wastes (Iles 2004, Pellow 2007). In contrast, this book analyzes the creation and development of the entire chemicals regime rather than focusing on individual cases or treaties. It thus seeks to contribute to understanding how different chemicals issues and policies are related.

Second, this book adds to the literature on science and politics in international environmental cooperation. Science shapes environmental politics, although the relationship between organized scientific work and policymaking is complex (P. Haas 1990, 2004; Jasanoff and Wynne 1998; Bäckstrand 2001; Parson 2003, Jasanoff and Martello 2004; Schroeder, King, and Tay 2008; Young 2008b). Scientific debates and assessments are at the forefront of many policy processes on chemicals, including the use of scientific advisory bodies (Kohler 2006). This book builds on scholarly work on knowledge creation and environmental assessments (Farrell and Jäger 2005, Mitchell et al. 2006) as it presents new empirical findings on how assessment information is generated and diffused across policy forums, affecting outcomes. Closely related to these issues, this study also contributes to the literature on precaution (O'Riordan and Cameron 1994, Sandin et al. 2002, Harremöes et al. 2002, Eckley and H. Selin 2004; Maguire and Ellis 2005; Whiteside 2006). Specifically, the book examines the role of the precautionary principle in international assessments and regulations of hazardous chemicals and how debates about the precautionary principle are carried across forums and shape policymaking.

Third, the book contributes to the literature on regimes and institutional linkages. Scholars have analyzed the creation, operation, and effectiveness of environmental regimes for decades (Krasner 1983; Young 1989, 1991a; Haas, Keohane, and Levy 1993; Levy, Young, and Zürn 1995; Victor, Raustiala, and Skolnikoff 1998; Joyner 1998; Wettestad 2001; Breitmeier, Young, and Zürn 2006; Young, King, and Schroeder 2008). Furthermore, analyses of institutional linkages examine political and functional characteristics and implications of related policy and management efforts within and across different geographical scales, forums, and instruments (Stokke 2001, Young 2002, H. Selin and VanDeveer 2003, Raustiala and Victor 2004, Oberthür and Gehring 2006; Gehring and Oberthür 2008; Chambers 2008). Much of this literature focuses on horizontal linkages, creating a need to pay more empirical and analytical attention to vertical linkages. This book analyzes both horizontal and vertical linkages. In doing so, it puts regime participants at the center of analysis, in contrast to the largely structurally oriented early literature on institutional linkages.

Fourth, related to the importance for scholars to study characters and implications of horizontal and vertical linkages, this book is part of a growing literature on multilevel governance. Much of the early literature on multilevel governance focused on policy developments within the European Union (EU) (Marks 1992, 1993; Hooghe 1996). Since then, multilevel governance analyses have been expanded to address cross scale policy developments in other regions, as well as connections among governance levels from the global to the local (Betsill and Bulkeley 2006; N. Selin and H. Selin 2006; Young, King, and Schroeder 2008; H. Selin and VanDeveer 2009). By analyzing the development of the chemicals regime, which is more complex than many other environmental regimes, this book contributes to our understanding of multilevel governance. While better-covered regimes structured around a framework convention have a relatively clear political center at the top, the chemicals regime lacks a single focal point. This study thus focuses on an empirical case where there is no well-defined apex of the governance structure, which creates particular governance challenges.

In short, the book demonstrates that policy developments on chemicals have frequently been driven by coalitions of states, IGOs, and NGOs formed around shared interests. These coalitions often engage in linkage politics. For example, many early policy developments focused on international trade issues and were pushed by a coalition of developing countries together with IGOs and NGOs seeking trade restrictions as a means to prevent foreign dumping of hazardous wastes and chemicals in their countries. In addition, a group of Northern countries in the early 1990s pushed issues of the long-range transport of hazardous chemical emissions onto the political agenda. Many of these actions targeted chemicals that had been phased out in most industrialized countries but were still in use in developing countries. This difference in focus between local management problems and transboundary pollution issues caused political tensions, but also created possibilities for bargaining and policy compromise across policy forums and regulatory instruments.

Institutional linkages frequently shape interests and actions of regime participants. Scientific and political debates and policy developments in one chemicals forum are not separate from those in another one simply because the forums are formally independent. In fact, a major characteristic of the chemicals regime is that coalitions of regime participants use agreements on particular rules and practices in one policy forum to leverage similar outcomes in another policy arena. This can, for example, be seen with respect to the diffusion of the principle of prior informed consent (PIC) addressing trade issues as well as the regulation of specific chemicals under multiple treaties. These policy diffusions are sometimes positive, as regime participants capture synergetic effects through the deliberate use of institutional linkages. In such cases, the strategic exploitation of institutional linkages facilitates decision making that results in complementary policy developments and management efforts across different parts of the regime.

Nevertheless, the effects and outcomes of institutional linkages are not always synergetic. The chemicals case also demonstrates that just as easily as institutional linkages can be used to diffuse complementary ideas and policies across forums, they can also be used to transfer political differences and struggles from one arena to another. In such instances, institutional linkages in effect make it more difficult to reach an agreement in one forum because unresolved conflicts are spilling over from a different one. This is, for example, the case with ongoing political debates across multiple forums about the establishment of more comprehensive mechanisms on monitoring, compliance, and capacity building. On these issues, coalitions of regime participants express diverging opinions as political stalemates permeate across policy forums. Related to these issues, participants in chemicals management are also struggling to better coordinate and link different management schemes toward improved governance.

Given these considerations, what are the major lessons from the chemicals case for other governance efforts? One major message is that increasing institutional density is likely to directly affect regime participants' interests and strategies. That is, states, IGOs, and NGOs not only think about their interests and strategies in the context of what is going on in one forum, but also how choices and actions in that forum will affect their interests and policy outcomes in other policy arenas and regime development efforts more broadly. It is also important to realize that institutional linkages are not automatically positive or negative. Whether institutional linkages will end up facilitating or hindering problem-solving efforts depends ultimately on how participants engaged in multiple forums view and elect to use those linkages. Finally, an increase in institutional density and linkages highlights the need to effectively link management efforts across global, regional, national, and local governance scales. All of these issues are discussed in the concluding chapter.

The remainder of this chapter gives a brief introduction to global chemicals policy and provides an overview of the structure of the book.

Global Chemicals Policy in Brief

The chemicals regime consists of a multitude of formally independent but functionally dependent treaties and programs. SAICM is designed to guide cooperative efforts and set policy objectives for enhanced environmental and human health protection, linking governance efforts across the chemicals regime. Three global treaties (the Basel Convention, the Rotterdam Convention, and the Stockholm Convention) and one regional treaty (the CLRTAP POPs Protocol) constitute the core of the chemicals regime, covering overlapping parts of the life cycle (see table 1.1). While other regional agreements do not receive the same amount of attention, the CLRTAP POPs Protocol is included even though it is regional in scope because of its importance in international chemicals management and its strong policy and management linkages to the three global treaties (in particular the creation and implementation of the Stockholm Convention).

In the life cycle management of hazardous chemicals, the Basel Convention regulates the transboundary movement and disposal of discarded or used chemicals if they fall under the treaty's definition of hazardous wastes. It furthermore recognizes the importance of waste minimization. The Basel Convention was adopted in 1989 and entered into force in

Table 1.1
Summary of the four main treaties on chemicals management

Basel Convention: Adopted in 1989; entry into force in 1992; 172 parties as of 2009	• Regulates the transboundary movement and disposal of hazardous wastes; covers chemicals if they fall under the treaty's definition of hazardous wastes. • Subjects hazardous waste transfers to a PIC procedure where an importing party must give explicit consent before shipment. • Prohibits exports of hazardous wastes to Antarctica and to parties that have taken domestic measures banning imports. • Exports of hazardous wastes to nonparties must be subject to an agreement at least as stringent as the Basel Convention. • The 1995 Ban Amendment (not yet in force) bans export of hazardous wastes from parties that are members of the OECD or the EU, as well as Liechtenstein, to other parties. • The 1999 Protocol on Liability and Compensation (not yet in force) identifies financial responsibilities in cases of waste transfer accidents. • Basel Convention regional centers address management and capacity-building issues.
Rotterdam Convention: adopted in 1998; entry into force in 2004; 128 parties as of 2009	• Regulates the international trade in pesticides and industrial chemicals using a PIC scheme. • Covered forty chemicals by 2009. • Requires an exporting party to receive prior consent from an importing party before exporting a regulated chemical. • Obligates parties to notify the secretariat when they ban or severely restrict a chemical. • Contains a mechanism for evaluating and regulating additional chemicals under the treaty.
CLRTAP POPs Protocol: Adopted in 1998; entry into force in 2003; 29 parties as of 2009	• Regulates the production and use of POP pesticides and industrial chemicals. • Outlines provisions for the environmentally sound transport and disposal of POPs stockpiles and wastes. • Sets technical standards for controlling emissions of by-product POPs. • Regulated sixteen chemicals by 2009. • Contains a mechanism for evaluating and regulating additional chemicals.
Stockholm Convention: Adopted in 2001; entry into force in 2004; 163 parties as of 2009	• Regulates the production, use, trade, and disposal of POP pesticides and industrial chemicals. • Sets technical standards for controlling the release of by-product POPs. • Contains a mechanism for evaluating and regulating additional chemicals. • Regulated twenty-one chemicals by 2009. • Stockholm Convention regional centers support capacity building and implementation.

1992. By late 2009, 171 countries and the EU were parties, making the convention one of the most widely ratified multilateral treaties in the world. The Basel Convention prohibits export of hazardous wastes to Antarctica and to parties that have taken domestic measures to ban imports. Hazardous waste transfers from one party to another are subject to a strict PIC procedure, under which a party must give explicit consent to a waste import before a shipment can take place. Exports of hazardous wastes to nonparties must also be subject to an agreement at least as stringent as the Basel Convention.

Basel Convention regulations have been strengthened over time. The Basel Ban Amendment, which prohibits export from Annex VII countries—members of the Organisation for Economic Cooperation and Development (OECD), EU countries, and Liechtenstein—to all other parties (mostly developing countries) was adopted in 1995. The 1999 Protocol on Liability and Compensation addresses who is financially responsible in instances of incidents and damages resulting from the transfer of hazardous waste covered by the convention. However, neither of these agreements has yet entered into force. Parties have also developed guidelines for the management of particular waste streams, including e-wastes, which in many cases contain hazardous chemicals. In addition, fourteen Basel Convention regional centers located in different parts of Latin and South America, Africa, Asia, and Europe have been established to aid implementation and capacity building.

The Rotterdam Convention focuses on the international trade in commercial industrial chemicals and pesticides. Based on an earlier voluntary PIC mechanism, the convention was adopted in 1998 and entered into force in 2004. By late 2009, 126 countries and the EU were parties. The Rotterdam Convention is designed principally to assist developing countries in deciding whether to permit the import of a specific chemical by increasing their access to information about hazardous chemicals that are subject to trade. Chemicals listed in the treaty can be exported from one party to another only after prior consent by the importing party. Parties are obligated to notify the convention's secretariat when they ban or severely restrict a chemical, so that information may be made available to other parties. The Rotterdam Convention also includes a mechanism for the evaluation and possible inclusion of additional chemicals; by 2009, the treaty covered forty chemicals.

The CLRTAP POPs Protocol operates under the auspices of the United Nations Economic Commission for Europe (UNECE), which comprises North America and Europe as far east as Russia and Kazakhstan. The agreement was signed in 1998 and entered into force in 2003. By late 2009, twenty-eight countries and the EU were parties. The protocol is designed to reduce the release and long-range transport of POP emissions. To this end, the protocol regulates the production and use of POP pesticides and industrial chemicals and controls the environmentally sound transport and disposal of POP stockpiles and wastes. It also sets technical standards and guidelines for controlling emissions of POPs that are generated as by-products. By 2009, sixteen chemicals were subject to regulations. The protocol also has a mechanism for evaluating additional chemicals for possible controls; several more POPs are on track to be regulated under it.

The Stockholm Convention targets the production, use, trade, and disposal of commercial POPs, as well as the release of POP by-products. The treaty was adopted in 2001 and entered into force in 2004. By 2009, 161 countries and the EU were parties. The Stockholm Convention regulates the production and use of POP pesticides and industrial chemicals. Parties are required to ban the import or export of controlled POPs except for purposes of environmentally sound disposal. On issues of the trade in discarded POPs and their disposal, the Stockholm Convention refers to the Basel Convention. Parties should also minimize releases of by-product POPs. By late 2009, the Stockholm Convention regulated twenty-one POPs, and several other chemicals are in the process of evaluation for possible controls through a treaty mechanism. Furthermore, parties are working to establish regional centers to support capacity building and implementation.

In addition, many regional agreements developed since the 1960s address hazardous chemicals in different ways. A large number of these are designed to protect shared seas and lakes against chemical pollution and dumping (including from many of the same chemicals that are covered by the four main treaties). Under the Regional Seas Programme run by UNEP, thirteen action plans targeting a long list of pollutants had been established by 2009, involving over 140 countries. In addition to these UNEP-led actions, regional agreements covering, for example, the Northeast Atlantic, the Baltic Sea, and the North American Great Lakes contain regulations on a wide range of hazardous chemicals. A growing number of treaties covering transboundary rivers all over the world also contain pollution-prevention measures. Furthermore, countries in several regions have created separate waste management agreements, sometimes in response to the Basel Convention.

Overview of the Book

The following seven chapters address a multitude of global governance issues with an empirical focus on the development of the chemicals regime. To this end, the book provides an in-depth analysis of key stakeholders and issues in the creation and implementation of the four major multilateral treaties that form the core of the chemicals regime: the Basel Convention, the Rotterdam Convention, the CLRTAP POPs Protocol, and the Stockholm Convention. The book also discusses how these four treaties relate to other agreements and programs that address hazardous chemicals in different ways. In analyzing the four major treaties, each treaty chapter explores key issues of policy expansions, institutional linkages, and the design of effective multilevel governance

Chapter 2 outlines a framework for analyzing multilevel governance and institutional linkages on chemicals. The chapter begins with a discussion of institutions and actors in regime analysis with a focus on the chemicals regime. This is followed by a discussion on important characteristics of multilevel governance and the influence of institutional linkages—separated into governance and actor linkages—on chemicals policymaking and management. It provides as well an overview of the main components of the chemicals regime, divided into principles, norms, rules, and decision-making procedures. This discussion relates specific regime components to issues of chemicals management, institutional linkages, and multilevel governance, which are examined in more detail in subsequent chapters. The chapter ends with a discussion about science and policy interplay issues under the chemicals regime.

Chapter 3 provides a historical perspective on the rise of scientific awareness and public concern about hazardous chemicals and the development of related policy and management efforts. The chapter begins by examining early action by states and IGOs on hazardous chemicals from the 1960s to the late 1980s, focusing on improving information gathering and the harmonization of domestic and international regulations of a few hazardous substances. This is followed by a discussion of the development of a more comprehensive chemicals policy, including the adoption of several multilateral treaties over the past two decades, culminating in ongoing efforts on improving the implementation and effectiveness of existing instruments, including through SAICM. Building on this discussion, chapters 4 to 7 focus on the four main international treaties addressing chemicals issues to date.

Chapter 4, which analyzes the Basel Convention, begins with a discussion of global waste trade issues, including chemicals wastes. This is followed by an analysis of the development and shift in the 1980s from a voluntary PIC policy on hazardous wastes to a mandatory procedure in the Basel Convention. This is continued by an examination of the implementation of the convention, highlighting several institutional linkages and multilevel governance issues, including the incorporation of the PIC principle for managing trade and related efforts to strengthen controls; the development of technical guidelines for waste management; the operation of regional centers for implementation and capacity building; and the establishment of mechanisms for liability, monitoring, and compliance. The chapter ends with a discussion of issues critical to the continued strengthening of hazardous waste management under the Basel Convention.

Chapter 5 examines the Rotterdam Convention. It starts by discussing policy and management issues relating to the trade in chemicals and then examines the establishment of a voluntary PIC procedure in the 1980s. Next, the chapter analyzes the decision by the international community to create the Rotterdam Convention and early implementation efforts. The chapter identifies several institutional linkages and multilevel governance issues, including the establishment of a PIC principle for managing trade, the creation of the Chemical Review Committee for evaluating additional chemicals for possible controls, the generation and structuring of financial and technical assistance, and efforts to develop mechanisms for monitoring and compliance. The chapter ends with a discussion of major challenges and opportunities for more effective management of the trade in hazardous chemicals under the Rotterdam Convention.

Chapter 6 analyzes the CLRTAP POPs Protocol. It begins with an examination of the rise of the POPs issues on the international agenda in the late 1980s, followed by an analysis of the CLRTAP scientific assessment work and the subsequent political negotiations resulting in the POPs protocol. Next, the chapter examines implementation efforts to date. It highlights several institutional linkages and multilevel governance issues, including the scientific and political framing of the POPs issue; the assessment and development of management options for specific POPs; and the creation of a review committee to evaluate additional chemicals that may be regulated under the protocol. The chapter ends with a few comments on major management issues related to the continuous implementation of the CLRTAP POPs Protocol and how these are linked to other instruments under the chemicals regime, including the Stockholm Convention.

Chapter 7 focuses on the Stockholm Convention. It begins by examining the elevation of the POPs issue from the regional level to the global level in the 1990s. It then analyzes the global POPs scientific assessment work and the negotiations of the Stockholm Convention. Next, the chapter examines early efforts on implementing the convention. It discusses several institutional linkages and multilevel governance issues, including the expansion of regional CLRTAP POPs controls to the global level; the formation of the Chemical Review Committee for the evaluation of additional chemicals; the creation and funding of organizational structures supporting capacity building; and the establishment of mechanisms for monitoring and compliance. The chapter ends with a discussion of major issues related to the continued implementation of the Stockholm Convention.

Finally, chapter 8 brings together main insights and arguments from earlier chapters on the development and implementation of the individual treaties. The chapter begins with a summary of the chemicals regime. Following this, it returns to the three research questions outlined in this chapter. Answering these research questions in turn, it highlights key roles of coalitions of states, IGOs, and NGOs in policymaking across policy forums, as well as how institutional linkages shape policy expansion efforts and multilevel governance. The next section identifies and discusses four governance issues important for improving the management of hazardous chemicals and fulfilling the goal set at the WSSD on achieving safe production and use of chemicals no later than 2020. The chapter—and the book—ends with a few concluding remarks on some major lessons that can be drawn from the chemicals case.

2

Institutional Analysis and the Chemicals Regime

Scholars have long studied how states, working with IGOs and NGOs, create, uphold, and expand joint structures addressing environmental issues (Krasner1983; Haas, Keohane, and Levy 1993; Levy, Young, and Zürn 1995; Victor, Raustiala, and Skolnikoff 1998; Breitmeier, Young, and Zürn 2006; Young, King, and Schroeder 2008). Regimes establish common standards for behavior and specify obligations that all parties within a particular issue area are expected to meet. As such, regimes can be important for enhancing transparency and increasing the political costs of particular environmentally destructive activities (DeSombre 2006). Recently institutional analysis has addressed issues related to linkages among different governance structures (Stokke 2001; Young 2002; H. Selin and VanDeveer 2003; Raustiala and Victor 2004; Oberthür and Gehring 2006; Chambers 2008; Young, King, and Schroeder 2008). As policies have been expanded within and across political forums addressing hazardous chemicals, the development and performance of one policy instrument or management effort may have significant impacts on policy processes and outcomes in other chemicals forums.

This study of the development of the chemicals regime draws heavily from the "new institutionalism" literature, which has become influential across multiple social science disciplines and scholarly fields (March and Olsen 1989). In short, this literature examines how social structures consisting of particular norms, rules, and decision-making practices shape actors' behavior and policy making. Within the broad literature on new institutionalism, this analysis shares many basic assumptions and approaches of researchers applying a "social-practices perspective" and a "knowledge-action perspective" (Young 2008a, 8–9). The social-practices perspective stresses that actors' interests and preferences are at least partly shaped through interaction and that social structures over time determine what is considered appropriate behavior. The knowledge-action perspective adds to this by arguing that leadership by actors together with social structures influence how environmental issues are understood and conceptualized, shaping discourses as well as policy processes and outcomes.

Combining arguments from the social-practices and the knowledge-action perspectives, this chapter presents a framework for analyzing the creation and implementation of the chemicals regime. This framework examines the relationship between structures and agents in international politics, drivers of policy expansions, and characteristics of institutional linkages and multilevel governance. The chapter begins with a discussion about institutions and participants in regime development, briefly describing the many roles of states, IGOs, and NGOs working on the management hazardous of chemicals. The next section addresses important characteristics of institutional linkages and multilevel governance issues, with a focus on policy making and management of hazardous chemicals. This is followed by brief summary of the principles, norms, rules, and decision-making procedures of the chemicals regime. The chapter ends with a discussion of regimes and science-policy interplay issues.

Institutions and Regime Participants

Institutional analysis requires a clear distinction between regimes that are social structures guiding the behavior of stakeholders but possess no independent ability to act, and the regime participants (states, IGOs, and NGOs) that make decisions and drive policy developments, thereby creating and upholding social structures (Young 1989, 2002, 2008a; H. Selin and VanDeveer 2003). Therefore, regime analysis should distinguish between structures and agents as well as empirically explore their relationships. In short, institutions function as guiding means as states, IGOs, and NGOs communicate with others, justify actions, and pursue goals. At the same time, regime participants establish, reinforce, and challenge institutions through their continuous behavior and interactions, as they frequently refer to broader social orders when making choices (Kratochwil 1989).

Institutions are socially created structures, or "persistent and connected sets of rules and practices that prescribe behavioral roles, constrain
activity, and shape expectations" (Keohane 1989, 3). In other words, institutions function as constituting and guiding social forces by defining
acceptable or legitimate behavior (Young 2002, 2008a, 2008b). In these
ways, international institutions regularly affect the behavior of regime
participants, policy outcomes, and management efforts. Regimes, as is-
sue-specific institutions, define acceptable behavior and shape perceptions
within a particular policy area. The process by which issues become ag-
gregated into a distinct issue area and the continuous development of that
issue area follows no predetermined pattern, but is an important aspect of
regime formation and operation (E. Haas 1980, Kratochwil 1993).

An international regime is commonly defined in the international relations literature as consisting of "sets of implicit and explicit principles,
norms, rules and decision-making procedures around which actors' expectations converge in a given area of international relations" (Krasner
1983, 2). In other words, a regime encompasses a particular set of issue-
specific rules and procedures that are embedded in associated principles
and norms, collectively defining encouraged and proscribed behavior
within a specific field of activity. In this respect, an international regime
is a socially constructed governance structure that shapes what is seen as
appropriate and inappropriate conduct in a specific policy area. A host
of regimes deals with environment and natural resource issues. In more
formalized regimes, many components are codified in one or several legal
instruments, as in the case of the chemicals regime.

Analysts use different methods to identify a regime. Raustiala and Victor (2004) discuss the existence of a "regime complex" on plant genetic
resources. They also briefly identify chemicals as another case where a
regime complex may exist, but they do not discuss any details. Although
their study and this book share many analytical approaches to studying
regimes and institutional density, there are conceptual differences. For
example, the regime definition that Raustiala and Victor offer is treaty
based; they define each legal agreement covering plant genetic resources as
an "elemental regime" and argue that two or more elemental regimes that
overlap in scope, subject, and time form a regime complex. In contrast,
this study uses more of an issue area–based definition of a regime. That
is, the chemicals regime is conceptualized as consisting of several treaties

and programs that are cognitively and practically linked by the fact that they focus on the life cycle management of hazardous chemicals.

When an issue area–based definition is used, regimes are seen to operate within discrete policy areas based on shared knowledge and understanding of a physical condition. However, issue areas are also socially defined: "Regimes as social configurations are not 'objective' like mountains or forests, but neither are they 'subjective' like dreams or flights of speculative fancy. They are, as most social scientists concede at the theoretical level, intersubjective constructions" (Hasenclever, Mayer, and Rittberger 1997, 42–43). Thus, an issue area consists of perceived linkages among interrelated sets of physical and social factors involving interdependence among issues and participants, where both issue and actor characteristics are of importance. Furthermore, as knowledge and regime participants' interests change, conceptualizations of issue areas can become outdated, leading to alterations in issue area content and boundaries.

A multitude of states, IGOs, and NGOs cooperate—and compete—under international environmental regimes. These repeated interactions over time influence their views of environmental problems as well as their interests and policy preferences. Almost all of the world's states, albeit to varying degrees, are engaged in chemicals *policy making* and management. Of course, national governments are the ones that formulate policy goals, negotiate treaties, and adopt management programs. In addition, states are responsible for the implementation of treaty obligations and programmatic commitments. States also collaborate through the many IGOs of which they are members. IGOs, like other kinds of organizations, are material entities with a capacity to act in social practices. They have physical locations, offices, personnel, equipment, and budgets (Young 1991a). Organizations act as arenas for political actions, serve as innovators, and, through their staff, participate in agenda setting, coalition formation, and negotiations (E. Haas 1990).

Many leading IGOs work on a host of legal, political, and technical issues under the chemicals regime (Alston 1978, Lönngren 1992, Asante-Duah and Nagy 1998). Each major chemicals treaty is administered by a secretariat that oversees and aids its implementation. The International Labour Organization (ILO) develops agreements on chemicals used in workplaces, including benzene, asbestos, and lead paint. The OECD coordinates testing requirements and establishes common guidelines for data generation and sharing among its member states. The United Nations Institute for Training and Research (UNITAR) works with developing countries to improve domestic capabilities for chemicals management. ILO, OECD, and UNITAR also collaborate on the implementation of the Globally Harmonized System for the Classification and Labeling of Chemicals, which facilitates identification of chemicals that are transported among countries.

UNEP is centrally involved in international chemicals politics and policy making, shaping the structures and content of all leading treaties and programs. It also provides support for SAICM and is active in the operation of a multitude of multilateral agreements that seek to protect oceans and regional seas from dumping and other types of pollution, including chemical pollution. On these issues, UNEP works with the International Maritime Organization. In addition, UNEP led the establishment of the International Register of Potentially Toxic Chemicals (IRPTC) in 1976 and continues to keep lists of hazardous chemicals. The United Nations Food and Agriculture Organization (FAO) and the World Health Organization (WHO) work on chemicals issues within their respective areas of expertise. They also collaborate on the Codex Alimentarius Commission, issuing recommendations to governments on acceptable levels of pesticide residues in foods.

The Intergovernmental Forum on Chemical Safety (IFCS) functions as a forum for debate among IGOs, national governments, industry groups, environmental NGOs, and scientific experts. In this capacity, it partakes in the implementation of SAICM. The Global Environment Facility (GEF), for which UNEP, the United Nations Development Programme (UNDP), and the World Bank are implementing agencies, work on several projects on chemicals management associated with the main international treaties. The Inter-Organization Programme for the Sound Management of Chemicals (IOMC) initiates and coordinates action toward fulfilling the WSSD 2020 goal for sound management of chemicals. To this end, the IOMC brings together the activities of seven different IGOs: FAO, ILO, OECD, UNEP, UNITAR, WHO, and the United Nations Industrial Development Organization (UNIDO).

Many environmental NGOs have long been active on chemicals policy making and management. NGOs can act as important catalysts of policy expansions by, for example, shaping debates, providing scientific information, developing policy proposals, and participating in transnational coalition building (Betsill and Corell 2008, Princen and Finger 1995). Notable NGOs include not only traditional organizations such as the WWF and Greenpeace, but also the Basel Action Network, the Pesticide Action Network, and the International Chemical Secretariat, which focus specifically on hazardous chemicals and wastes. In addition to these environmental NGOs, Arctic indigenous peoples groups have been a major presence in the development of international chemicals policy since the 1990s. The Inuit Circumpolar Council (ICC), representing Arctic indigenous communities, is particularly active on POP issues in regional and global forums (N. Selin and H. Selin 2008).

Given the economic power of the chemicals industry, the size of the international trade in chemicals, and the strong trade and market dimensions of chemicals management, it is hardly surprising that major industry associations and leading multinational firms are active participants in the development of domestic and international chemicals policy (Garcia-Johnson 2000; H. Selin 2000, 2007). The International Council of Chemical Associations (ICCA), the American Chemistry Council (ACC), and the European Chemical Industry Council (CEFIC) are influential industry organizations that together with many firms regularly attend international scientific and political meetings, including COPs and subsidiary bodies meeting under chemicals treaties. Private sector representatives are also in frequent contact with state officials or are included in national delegations to international meetings as they try to shape policy decisions. In particular, private sector actors focus on the controls of specific chemicals and the design of trade restrictions.

Industry organizations and firms also work on voluntary programs that operate alongside multilateral treaties and national law. In many cases, organizations and firms have promoted voluntary programs as an alternative to expanding mandatory requirements. This can, for example, be seen with respect to the management of particular waste streams under the Basel Convention and the Responsible Care program. The Responsible Care program, which was started in Canada in 1985 and currently has members in over fifty countries, is managed by the ICCA in collaboration with national associations. The program seeks to improve the safety and environmental performance of chemicals firms and their products. Although so-called type II partnerships—voluntary agreements between public and private sector entities— have been promoted at the WSSD and elsewhere as an innovative way to address environmental and sustainable development issues, many NGOs and states express concerns about relying too heavily on voluntary commitments and partnerships for risk management.

Across issue areas, leadership by coalitions of or individual states, IGOs, and NGOs can be critical to regime development (Young 1991a, Underdal 1994, H. Selin and VanDeveer 2003, Skodvin and Andresen 2006). Groups of individual experts from states and organizations— epistemic communities—may also shape policy making (P. Haas 1990, 2004). Sometimes part of linkage politics, leadership can be exercised in many forms, including intellectual and structural (Young 1991b). An intellectual leader relies on the power of ideas, norms, and knowledge to shape the way other participants involved in regime formation perceive issues and conceptualize policy alternatives. Intellectual leaders often seek the adoption of particular policies by trying to secure acceptance of new ideas, norms, and knowledge. Structural leadership, which can also be characterized as material leadership, is exercised through the commitment of financial, technical, and scientific resources necessary for environmental assessment and policy making, with the intent of shaping agendas and policy outcomes.

Of course, intellectual and material leadership efforts can be combined; the two forms of leadership are not mutually exclusive. In fact, it is a relatively common tactic for environmental leader states, in collaboration with IGOs and NGOs, to combine intellectual leadership with different kinds of material support for policy change (Liefferink and Anderson 1998, H. Selin and VanDeveer 2003, Börzel 2002, H. Selin 2007). Such combined leadership is often exercised by advocating for particular policy expansions in combination with the willingness to host and finance international political and scientific meetings, sponsor international environmental scientific assessments, and fund national experts in international organizations preparing background documents and policy proposals. In such cases, policy leaders use combinations of intellectual and structural leadership to gain greater influence over policy-making processes than if they had relied on only one kind of leadership strategy.

Coalitions of states, IGOs, and NGOs have played many leadership roles on chemicals policy making. Specific policy issues were put on political agendas at different times as a result of intellectual and material leadership provided by groups of regime participants. Involving linkage politics, these leadership efforts often include "forum shopping," as policy advocates seek to find a suitable forum receptive to their concerns, as well as strategically using one forum to influence debates and decisions in a related forum (H. Selin and VanDeveer 2003, Raustiala and Victor 2004). The chemicals case also illustrates a less commonly analyzed activity, "scale shopping" (Young 2008a, Gupta 2008). That is, regime participants sometimes seek to scale up or scale down where an issue is addressed because they see practical benefits of (at least temporarily) dealing with an issue at one governance level over another. Many leadership, forum shopping, and scale shopping issues also involve science-policy linkages, which are discussed below.

Multilevel Governance and Institutional Linkages

Multilevel governance, involving linkage politics, intellectual and structural leadership, forum shopping, and scale shopping, is characterized by a multitude of horizontal and vertical linkages between and across geographical scales: global, regional, national, and local. Oberthür and Gehring (2006) identify a number of similar terms in the literature denoting instances of one instrument influencing another, including *linkage, interplay, interlinkage, overlap,* and *interconnection.* Here, the term *linkage* is primarily used, but the different terms are more or less synonymous. Because global chemicals governance is developed in multiple centers across geographical scales, and policy-making and regulatory authority is dispersed among legal instruments and political forums, implementation of the major chemicals treaties increasingly intersects. This has resulted in a number of important governance and actor linkages across different instruments and forums within the chemicals regime.

Institutional horizontal and vertical linkages may have both synergetic and conflicting effects on governance (Rosendal 2001). In a synergetic situation, separate policy instruments and management efforts addressing related issues are mutually supportive and reinforcing. Furthermore, synergy is often defined as enhancing effectiveness, which aids management and collective problem solving. That is, the outcome in a synergetic situation is more than the mere sum of the different parts; value is added in that the total result is greater than what could have been achieved by each piece working separately. In contrast, conflicting effects arise when the objectives or operation of two or more interrelated policy instruments and management efforts contradict each other. Then the separate policy instruments and management efforts negatively affect each other, hampering collective efforts to address environmental and human health problems.

In addition, there may be competition between policy venues as stakeholders in different forums compete for influence beyond the boundaries of those forums to become standard setters and policy leaders within entire issue areas (H. Selin and VanDeveer 2003). In cases of such linkage politics and venue competition, policy developments in one forum can act as an incentive for expanded policy making in another forum so as not to fall behind the first forum. This may lead to interrelated policy expansions across multiple policy venues. Such policy expansions can increase synergetic or conflicting effects across venues depending on the complementary and contradictory character of the sets of policies that result from venue competition. Furthermore, a push toward higher standards across related political venues may not benefit both venues equally: regulatory competition may enhance the political and regulatory status of one forum or instrument, diminishing the status of regulatory efforts in another forum.

Rothstein (2003) proposes that the more a specific policy issue is spread across different jurisdictions and involves multiple sets of stakeholders, the higher the likelihood is of the development of conflicting regulatory cultures and policy outcomes across policy venues. In contrast, when Oberthür and Gehring (2006) compared instances of synergetic and disruptive interactions across sets of related global and regional legal instruments, they found synergetic effects more prevalent than disruptive outcomes within the field of international environmental politics. That is, they showed that governance efforts across the different forums they included in their comparative study were more likely to be enhanced than impeded by linkages. Building on their findings and other empirical studies of institutional linkages, this book explores characters and implications of synergetic and conflicting linkages across chemicals forums and governance scales.

Closely related to the analytical need for clearly differentiating between structures and agents when studying regimes, it is important to distinguish between governance and actor linkages in institutional and multilevel analysis (H. Selin and VanDeveer 2003). Governance linkages refer to structural, immaterial connections in the form of related principles, norms, rules, and decision-making procedures across regulatory instruments. Because of the growing importance of governance linkages, the development and operation of nominally separate instruments often intersect on a de facto basis, having an impact on policy making and implementation (Young 1996). Actor linkages are agent-based linkages across separate instruments, where representatives of states, IGOs, and NGOs operate in two or more related forums. In many such cases, overlapping membership and the participation of similar sets of states, IGOs, and NGOs can have significant effects on behavior and outcomes across forums.

States typically strive to coordinate policy positions and outcomes across forums to ensure complementary policy goals and the design of supportive management efforts. However, different government agencies of a state or organizational subdivisions can, either because they are unaware of each other's actions or have different interests, pursue conflicting ends in different forums. If two forums on chemicals management adopt conflicting goals and regulations, members to both are faced with a dilemma. Problems can also arise if all but a few states are members of two related forums and want to combine activities. For example, in attempts at improved global chemical management, some states are parties to all major instruments, while others are not. This situation poses a continuing challenge for SAICM and other coordination efforts to enhance effectiveness and implementation under the chemicals regime.

Because the same individual representatives of states, IGOs, and NGOs often participate in multiple forums, many leading individuals are engaged in policy developments across multiple treaties and programs. Individuals play crucial roles in international cooperation by participating in coalition formation, agenda setting, decision making, and deal brokering, in these capacities regularly affecting outcomes under the chemicals regime. Furthermore, people may influence multiple organizations by changing professional affiliations and jobs over time. It is not unusual for one person to work in different positions and organizations within the same regime for several years or decades, building important personal networks within and across related forums. In addition, as people move between jobs and organizations, they take their experience, knowledge, ideas, and personal contacts with them.

Institutional linkages may affect both decision-making processes across policy forums and the effectiveness of related policy instruments. Oberthür and Gehring (2006) identify four causal mechanisms of institutional interactions that can be used to analyze how actions and decisions in one policy forum affect choices and management efforts in another forum. The first two mechanisms focus on decision-making issues, and the second two mechanisms address effectiveness issues. In cases of *cognitive interaction,* stakeholders' transference of ideas and knowledge, including technical and scientific concepts, practices, models, and reports, shapes decisions in multiple policy forums. When there is *interaction through commitment,* parties' acceptance of particular principles, norms, and rules in one policy forum affect stakeholders' interests and decision making in another policy forum.

With respect to outcome-related linkages, in cases of *behavioral interaction,* changes in activities in one policy forum affect the effectiveness of other forums. This kind of interaction occurs, for example, when participants in one forum create political or financial incentives for states or nonstate actors to behave in a particular way. The actions taken in response to these incentives may then also shape the effectiveness of external governance efforts. Finally, in cases of *impact-level interaction,* the ultimate governance target of one policy instrument is affected by spillover effects from particular decisions or actions taken under another, functionally interdependent instrument. This book draws on this typology of four kinds of causal mechanisms to examine how policy decisions and management actions taken in one policy forum under the chemicals regime affect policy making and implementation in a related one.

Components of the Chemicals Regime

Like any other international environmental regime, the chemicals regime consists of several components established to guide the actions of states and other stakeholders working to mitigate environmental and human health risks of hazardous chemicals. Following a long tradition in regime analysis, these components can be broken down into a set of principles, norms, rules, and decision-making procedures.

Principles are fundamental standards that guide interactions in the international system and across policy areas and regimes. For example, the principle of common but differentiated responsibilities has been advanced in numerous areas of North-South environmental and economic relations. This principle stipulates that industrialized countries often have an obligation to take more ambitious regulatory action initially as developing countries are given more time to catch up (but are still expected to act) (Najam 2005). The idea of common but differentiated responsibilities is also frequently invoked in discussions about technical and financial assistance, including under the chemicals regime. In fact, capacity building has emerged as a central political linkage issue in the development of the chemicals regime. This study examines how the principle of common but differentiated responsibilities has been applied and contested within the chemicals regime and how these controversies shape regulatory and capacity-building efforts across instruments and programs.

The PIC principle, the polluter pays principle, and the precautionary principle are three other principles important in much international environmental policy making, including the creation and implementation of the chemicals regime. In short, the PIC principle in environmental politics outlines a shared responsibility between the exporter and importer of a particular good to address trade-related concerns and regulations. This principle originated within the chemicals area and has since been transferred into other issues areas such as the biodiversity regime (Wolf 2000). The development and diffusion of the PIC principle is critical to understanding politics and policy making in many parts of the chemicals regime. The transformation from an early voluntary PIC procedure for dealing with the international trade in hazardous chemicals and wastes to legally binding PIC procedures under the Rotterdam Convention and Basel Convention, respectively, is a major policy development involving much linkage politics under the chemicals regime.

The polluter-pays principle guides much international environmental policy making and management with respect to who has a responsibility to take regulatory action and fund abatement and cleanup activities. The precautionary principle has become an important principle guiding decision making and regulation in cases of scientific uncertainty in numerous environmental policy fields. This principle, as defined in principle 15 of the Rio Declaration adopted at the United Nations Conference on Environment and Development in 1992, states that "where there are threats of serious or irreversible damage, lack of full scientific certainty shall not be used as a reason for postponing cost-effective measures to prevent environmental degradation." Since the late 1980s, the precautionary principle—or language recognizing the importance of precaution– has been included in a large number of multilateral agreements and statements on hazardous chemicals (Eckley and H. Selin 2004, Whiteside 2006).

Operating alongside the regime principles, norms are explicit or tacit moral standards of appropriate behavior in specific situations. Like principles, they emerge and may become influential across issue areas. Many specific norms can be important in international cooperation as regime participants internalize them and act accordingly, more or less by habit. For example, most states share a normative belief that many transboundary issues are most appropriately dealt with multilaterally. As a result, the chemicals regime focuses on several issues that are transboundary in one way or another, including the international trade in commercial chemicals and chemicals wastes and the long-range transport of emissions. Norms related to the transfer of resources and capacity-building issues are also embedded in the regime. As such, this study examines the emergence and development of normative reasons for expanding multilateral cooperation and policy making on chemicals across separate policy forums.

Rules and decision-making procedures are exclusively connected to a specific issue area. They are typically negotiated and outlined in written multilateral agreements in more formalized regimes such as the chemicals regime (Krasner 1983, Young 1989). Rules are mandatory stipulations of what parties should and should not do in connection with the issues and activities covered by the regime. The chemicals regime relies heavily on legally binding rules that have been negotiated under multiple parts of the chemicals regime, including requirements to regulate the production, use, trade, and disposal of commercial substances. Treaties also set out stipulations to reduce emissions of particular by-products and require parties to conduct regular implementation reviews. As a result, the effectiveness of the chemicals regime is closely tied to the successful implementation of these rules across all major agreements.

Not all rules, however, are laid out in an equally detailed way in treaties (Victor, Raustiala, and Skolnikoff 1998). Like many other multilateral environmental agreements, leading chemicals agreements address several important implementation issues—including those pertaining to reporting, monitoring, compliance, and liability—only more broadly. These kinds of rules can be important for improving transparency and regime effectiveness. A regime can enhance transparency by, for example, subjecting states to public reporting requirements and establishing monitoring procedures. While it is rare to find strong monitoring and compliance programs in environmental regimes, states have sometimes been willing to accept more independent monitoring efforts to improve the availability of information and enhance implementation. Similarly, enforcement mechanisms are relatively weak in most environmental regimes, but treaties can set collective guidelines and requirements for state actions.

While the chemicals treaties stress the importance of mechanisms addressing issues of reporting, monitoring, compliance, and liability, they largely leave the specifics of the development of these mechanisms to the conferences of the parties (COPs). However, negotiations to establish such mechanisms have been difficult. In fact, some of the most contentious debates at many recent COPs have concerned these issues. In addition, there are strong cognitive and practical connections between how these issues are handled across policy venues and instruments. This book examines how reporting, monitoring, compliance, and liability issues are attracting growing attention and controversy in multiple forums under the chemicals regime and how these issues are related to efforts to improve regime effectiveness. In this respect, the chemicals regime reflects the status of these debates within much international environmental governance.

Finally, decision-making procedures are prevailing practices for making and implementing collective rules and decisions. The decision-making procedures for individual treaties are initially established by consensus during the intergovernmental negotiations leading to treaty adoption. Most treaty bodies also operate on the basis of consensus. Subsequent reviewing and revising of regime provisions and the development of management programs are left to the COPs of each treaty. In addition, science advisory bodies in the form of free-standing chemical review committees have been set up under chemicals treaties to review additional substances for possible regulation in accordance with criteria and procedures outlined in each treaty. The operation of the COPs and subsidiary bodies and how decisions are made are therefore critical to continuous regime development.

The Chemicals Regime and Science-Policy Interplay

One particular area where institutions, including environmental regimes, can be influential is with respect to the generation and dissemination of scientific information (Lidskog and Sundqvist 2002). As in many other environmental issue areas, the management of hazardous chemicals draws heavily from scientific data and assessments. This raises important issues about principles and procedures for the production and application of "usable knowledge" for policy making (P. Haas 2004). The interplay between science and policy is central in the chemicals regime as well as an important topic in regime analysis. Specifically, the way in which scientific assessment bodies are established and how they formulate scientific advice with the intention of influencing decision making can be controversial. This can, for example, be seen in the assessments by the chemical review committees under several of the main treaties as their work and the regulation of chemicals are influenced by institutional linkages.

One notable management problem is that policy making on issues such as hazardous chemicals that are characterized by high degrees of risk and uncertainty runs the chance of running into a legislative standstill because of paralyzing scientific and political controversy. Such inaction may inadvertently result in much harm to the environment and human health. A growing number of studies conclude that much environmental and human health damage could have been avoided by a quicker policy response to past early warnings about significant risks under conditions of early scientific uncertainty (Eckley and H. Selin 2004, Schörling and Lund 2004). Furthermore, a comparative analysis of several historical cases of risk regulation demonstrates that policy makers in the past have been more likely not to regulate something that later turned out to be harmful than to err on the side of caution and regulate something despite uncertainty about risks to the environment and human health (Harremöes et al. 2002).

Many analysts and policy makers argue that the precautionary principle offers one major (if not necessarily perfect) way to address problems of undesirable regulatory inaction and avoid unnecessary harm to the environment and human health (European Commission 2000, Sandin et al. 2002, Schörling 2003, Whiteside 2006). Dominating formulations of the precautionary principle stipulate that where there are threats of serious or irreversible damage, lack of full scientific certainty should not be used as a reason for postponing cost-effective measures to prevent environmental degradation (O'Riordan and Cameron 1994, Eckley and H. Selin 2004). From this perspective, advocates of the precautionary principle have criticized traditional methods of risk assessment, particularly those based on approaches that emphasize scientific certainty, for being too reactive and resulting in undesirable harm to the environment and human health (Raffensperger and Tickner 1999, Harremöes et al. 2002).

However, the role of precaution in international environmental politics is contested (Eckley and H. Selin 2004; Whiteside 2006; H. Selin 2009a, 2009b). In particular the EU and leading environmental NGOs have championed the inclusion of the precautionary principle in a multitude of multilateral agreements and argue that its application is essential for effective environmental and human health protection from hazardous chemicals. In contrast, the United States and the chemicals industry have preferred treaty language stressing the importance of scientific evidence before regulatory action is taken and have opposed including provisions explicitly referring to the precautionary principle. In addition, the EU, the United States, and other countries have sometimes clashed with respect to the design of assessment procedures and possible regulation of specific chemicals based on their different attitudes toward the incorporation of precaution in risk assessment and decision-making processes across policy forums.

Issues of precaution are closely related to the use of organized scientific assessments, which have become common in many environmental issue areas. "Good" chemicals policy is often described as grounded in "sound science." Scientific environmental assessments are highly dynamic social processes "by which expert knowledge related to a policy problem is organized, evaluated, integrated and, presented in documents to inform policy choices or other decisionmaking" (Farrell and Jäger 2005, 1). Assessment processes and reports are often critical for bringing attention to environmental conditions and acquiring policy-relevant information, as it is difficult to make meaningful progress without a basic scientific consensus on the nature of the core problem among leading regime participants (Kohler 2006). Furthermore, the implementation of the many chemicals treaties, including regulations on additional substances, depends on periodic scientific reassessments of related environmental and human health conditions (P. Haas 1990, H. Selin and Eckely 2003).

Environmental knowledge and policy are coproduced in processes of joint evolution from scientific findings through national and international acceptance of causality to their use in policy making (Jasanoff and Wynne 1998). Because assessments require specialized technology and human skills, the unequal global distribution of such capabilities involves a risk of biases in favor of states that have greater access to relevant scientific, technical, financial, and human resources. In combination with a willingness to invest these resources for knowledge creation and knowledge sharing, this can have significant effects on policy outcomes: it can give particular states and groups the ability to shape agendas and fashioning agreements suited to their needs and interests. Science thereby can become "politics by other means" (Harding 1991, 10). Linkage politics in this area may also involve forum-shopping and scale-shopping behavior by states pursuing particular policy measures across governance scales.

For an assessment to provide usable knowledge and be influential, the scientific process and the resultant report should be broadly regarded as scientifically credible, politically legitimate, and policy salient (Mitchell et al. 2006). Scientific credibility reflects the scientific believability of a report within the scientific community writ large. Political legitimacy is a measure of the political acceptability and perceived fairness of the assessment process and report to all participants. Policy salience reflects the ability of an assessment to address the end users' particular policy concerns so that they find the report useful. However, participants within and across policy forums sometimes express different views on what information they deem to be scientifically credible, politically legitimate, and policy salient. Stakeholders also frequently cite particular scientific reports and claims that support their own position while downplaying less supportive findings (Susskind 1994).

Participants in a host of environmental regimes have established subsidiary science advisory bodies. The organization and mandates of these advisory bodies differ greatly. Their designs range from loose ad hoc arrangements with few formalized instructions and irregular meeting schedules to standing committees with detailed rules for participation and work procedures that meet at set intervals (Farrell and Jäger 2005, Mitchell et al. 2006). The design of the science advisory bodies in the form of chemical review committees associated with the major chemicals treaties are toward the latter end of this spectrum. These operate formally independently under the Rotterdam Convention, the CLRTAP POPs Protocol, and the Stockholm Convention. The review committees are tasked with assessing substances nominated by parties for possible regulation. These assessments, together with a recommendation regarding controls, are forwarded to the COPs for decision making.

The creation and operation of the chemical review committees raise critical issues regarding their mandates, membership, and role in expanding treaty controls on chemicals (Kohler 2006). Regarding their mandates, there are detailed procedures for each step of the review process, including the kind of data needed for assessment. Membership for the Rotterdam and Stockholm conventions review committees is restricted, while the task force overseeing the CLRTAP review work is open to all parties. For the Rotterdam and the Stockholm conventions review committees, negotiations on membership were contentious before parties agreed on formulas based on geographical representation. Committee members' disciplinary and professional backgrounds are also important. As the different review processes are structured so that new chemicals should be regulated only if they meet certain scientific criteria, the assessment work by the review committees and related political debates at the COPs require close attention.

3

Global Chemicals Use and Management in a Historical Perspective

In an analysis of the development of the chemicals regime, the connection between the advancement of scientific understanding, the emergence of public apprehension, and the formulation of policy responses needs to be considered from a historical perspective. This chapter begins with an introduction to key chemicals management issues, followed by an overview of the rise of scientific and public consciousness of hazardous chemicals starting in the 1960s. Early concerns about growing chemicals use focused on risks to human well-being and local environmental damages from the application of pesticides and the mismanagement of industrial chemicals. As scientific knowledge of the characteristics and environmental behavior of hazardous chemicals improved in the 1970s and 1980s, policy makers and environmental advocacy groups expressed additional concerns about the long-range transport of emissions and adverse effects on humans and wildlife.

The overview of developments in scientific and policy awareness about hazardous chemicals is followed by a discussion of the roles that leading organizations and states played in the creation of international policy programs and agreements, including how these initiatives relate to advancements in scientific understanding and public concern. This is followed by a presentation of early domestic and international action on hazardous chemicals from the 1960s until the late 1980s. Many of these policy efforts focused on information gathering and the development of limited domestic and international regulations on a few hazardous substances for which there were early scientific data. This presentation is continued by an outline of the initiation of a more comprehensive life cycle chemicals policy since the 1990s, including continuing efforts to improve implementation across the main parts of the chemicals regime and enhance multilevel governance.

Chemicals Management Issues

The famous DuPont slogan, "Better Things for Better Living … Through Chemistry," from the 1930s captured the early optimism of the chemicals revolution, as production and use increased sharply following World War II.[1] A wide range of chemicals helps to improve human standards of living in a host of ways, including increasing yields of major cash crops, advancing public health protection from vector-borne diseases, and producing countless industrial and consumer goods. The chemicals industry consists of the many firms that produce chemicals from raw materials (mainly petroleum), as well as those that alter or blend individual substances into different mixtures. Some chemicals are produced in volumes of millions of metric tons per year, but most are produced in quantities of less than 1,000 metric tons annually (OECD 2001). There is, however, no easily accessible figure for the total amount of global chemicals production.

Over 100,000 chemicals have been registered in the EU for commercial use since the 1960s. The European Commission (2001) estimates that 40,000 to 60,000 chemicals are currently sold on the EU market. Most of these chemicals are likely also used in many other regions. Global sales of chemicals grew almost ninefold between 1970 and 2000. The OECD estimated in 2001 that the global output of the chemical industry will roughly double between 1995 and 2020 (OECD 2001). Worldwide chemical sales (excluding pharmaceuticals) are worth approximately $2 trillion (CEFIC 2006), constituting close to 10 percent of all global trade (OECD 2001). Asia (mainly Japan, China, and India) is the world's leading chemicals-producing region in monetary terms, closely followed by the EU and the United States. The market is dominated by a few dozen multinational companies, including BASF, Bayer, Dow Chemical Company, Shell Chemical, and DuPont (Datamonitor 2005, 2006).

Chemicals are released through agricultural activities, industrial production, combustion processes, and leakages from waste streams. Hazardous chemicals exhibit several important properties (see box 3.1). Emissions can travel long distances from their sources on air currents and waterways and in migratory animals. Some hazardous chemicals bioaccumulate, or build up in the fatty tissues of individual organisms, and further biomagnify upward through food chains. Environmental risks include estrogenic effects, disruption of endocrine functions, impairment of immune systems, functional and physiological effects on reproduction capabilities, and reduced survival and growth of offspring. Human long-term, low-dose exposure has been linked to carcinogenic and tumorigenic effects as well as endocrine disruption as authorities in many countries express concerns about such chemical exposure.

Chemicals poisoning occurs all over the world. Symptoms include a wide range of effects, from burning eyes, headache, and muscle cramps to loss of consciousness and death. National governments have regulated the production and use of specific chemicals since the 1960s. Chemicals management involves balancing conflicting interests. In regulating chemicals,

Box 3.1
Characteristics of hazardous chemicals

Persistence: The more persistent a chemical is, the longer it remains in the environment before it biodegrades. Persistence can be measured in several ways, including half-life (the amount of time it takes for a chemical to decay to half its value) in air, soil, water, and sediment. More persistent chemicals have longer half-lives. Persistence itself is not dangerous, but it gives rise for concern if a chemical exhibits other undesirable qualities with respect to toxicity, bioaccumulation, and biomagnification.

Toxicity: A chemical's toxicity can be measured by focusing on chemical, biological, and physical entities. Toxicity refers to the effect a chemical may have on an organism or part of an organism (organ, tissue, or cell). Different organisms may exhibit different responses to the same dose of a toxic chemical. Common concerns about toxic chemicals include their ability to cause different kinds of cancer, act as endocrine disrupters, and negatively affect human development in early developmental stages.

Bioaccumulation: An essential biological process that takes place in all living organisms to obtain necessary nutrients. Problems can arise when hazardous substances that have been released into the environment are accumulated through the same mechanism as the nutrients, allowing them to build up in fatty tissues of organisms over time. Consequently organisms accumulate higher body burdens of many hazardous chemicals as they age.

Biomagnification: A biological process related to bioaccumulation as hazardous chemicals that have bioaccumulated in a large number of organisms at a lower trophic level are concentrated further by an organism at a higher trophic level as those chemical concentrations are passed up food webs. As a result, species at the top of food webs (including humans) typically have higher concentrations of hazardous chemicals in their bodies than species lower in the same food web.

Source: H. Selin (2009b)

policy makers and regulators attempt to maintain a socially acceptable level of environmental and human health protection while not unnecessarily restricting benefits of modern chemistry. In this respect, chemicals management presents decision makers and regulators "with some of the most intractable dilemmas of social regulation" (Brickman, Jasanoff, and Ilge 1985, 21). Many important regulatory decisions must furthermore be taken in the face of scientific and socioeconomic uncertainties due to a lack of adequate assessment information (European Commission 2001, Eckley and H. Selin 2004).

The Chemicals Revolution

Of the many chemicals that have been regularly used since the beginning of the chemicals revolution, some were originally synthesized in the late nineteenth century. Widespread commercial introduction of chemicals began as early as the 1920s. The respective histories of PCBs and DDT (dichlorodiphenyl trichloroethane) illustrate important developments in the early use of modern chemicals and the subsequent rise in scientific and public concern of growing chemicals exposure.

PCBs are mixtures of chlorinated hydrocarbons that were first developed in 1881, although it was not until 1929 that the American company Monsanto introduced them to the commercial market. Production and use of PCBs increased rapidly after World War II, and by the 1970s PCBs were widespread in a variety of industrial products and consumer goods worldwide (Koppe and Keys 2002). Because of their insulating capacity and effectiveness as flame retardants, PCBs were extensively used in transformers and other types of electrical equipment that were an integral part of post–World War II industrial and economic development. PCBs were also used in hydraulic and lubricating oils, paints, lacquers, and varnishes and as pigments in various plastics. All of these areas of application contributed to the economic importance of PCBs, and their frequent use was not initially seen as problematic (Eckley and Selin 2004).

It is believed that an Austrian student in the course of his thesis work was the first to create DDT in 1873 (Mellanby 1992). However, he did not realize its potential as a pesticide and thus failed to reap any significant professional or economic benefit from his discovery. Instead, the development of DDT as a pesticide was the result of the decision of a

Swiss chemical company, Geigy, in 1935 to expand its activities into the new and rapidly growing market for pest control. The company's aim was to synthesize a new pesticide that was easy and cheap to produce, had a wide area of agricultural application, exhibited long-lasting effects on crop-damaging insects, and was harmless to sprayed plants as well as local animals and humans. At the time, no existing substance had all these characteristics, and the development of such a pesticide held great profit potential.

Geigy scientists in 1939 came up with DDT, which seemed to fulfill all the set criteria. The outbreak of World War II delayed the commercial introduction of DDT, but DDT was used during the war, primarily by the Allied forces, to prevent insect-borne tropical diseases such as typhus, bubonic plague, and malaria. As a result of its wartime uses, DDT gained a reputation as a wonder drug. When it became available for civil use shortly after World War II, it became widely used in agriculture. Pest control was its main area of application in the 1950 and 1960s, although large quantities were also used against malaria mosquitoes for human health purposes in tropical and subtropical climates (Mellanby 1992). Paul Müller, the leading chemist at Geigy working on DDT, received the Nobel Prize for Medicine and Physiology for the discovery of DDT's pesticide properties in 1948. The most euphoric accounts of DDT's benefits claimed that most pests would soon be permanently wiped out.

The so-called Green Revolution program initiated by the Rockefeller Foundation in the 1940s, which fueled much pesticide use, was aimed at developing new high-yield varieties of major cash crops such as rice, wheat, and maize to meet increased demands for food from a growing global population. The Green Revolution program was implemented in the shadow of the Cold War and was in part used by U.S. interests as a way to provide more food to prevent social unrest and the spread of communism in developing countries. It significantly contributed to higher crop yields, in conjunction with the design of better irrigation schemes and increased use of nitrogen fertilizers (Linnér 2003). Many new high-yield plant varieties, however, were susceptible to different kinds of diseases and pests, and thus required the application of large amounts of toxic pesticides to produce maximum yields.

A small group of scientists and concerned citizens voiced skepticism about the indiscriminate application of DDT and similar kinds of new pesticides as early as the 1940s (Brooks 1972). However, these concerns were more or less ignored by the larger scientific and policy communities. Instead, it was largely because of the groundbreaking work by Rachel Carson, beginning in the late 1950s and publicized in her book *Silent Spring* in 1962, that issues of unregulated chemicals use and associated environmental and human health effects became publicly debated and politicized. Carson, who initially had difficulty finding a publisher for her book, warned that the widespread use of DDT and similar chemicals such as chlordane, endrin, and dieldrin contributed to the extinction of local bird populations—causing silent springs—and exposed other animals and people to great chemical hazards.

In 1972, Paul Brooks (1972, 222) observed that "within a decade of its publication *Silent Spring* has been recognized throughout the world as one of those rare books that change the course of history—not through incitement to war or violent revolution, but by altering the direction of man's thinking." Before *Silent Spring* was published in fall 1962, it appeared as a series of articles in the *New Yorker* over the previous summer. This serialization in the *New Yorker*, which helped Carson find a publisher for her manuscript, garnered great public interest in the book and the concerns it raised. Carson's claims, however, were also met by much skepticism and even outright hostility from several groups (Brooks 1972, Lear 1997). Brooks (1972, 293) noted further that "perhaps not since the classic controversy over Charles Darwin's *The Origin of Species* just over a century earlier had a single book been more bitterly attacked by those who felt their interests threatened."

Some of Carson's strongest critics were in the chemicals industry and the U.S. Department of Agriculture. Attacking both her person and her work, they accused Carlson of scaremongering, dismissing her arguments as "science fiction" (Lytle 2007). A former secretary of agriculture, Ezra Taft Benson, is commonly credited with asking, "Why is a spinster with no children so concerned about genetics?" His answer: Carson was "probably a communist"! Carson, however, gained growing support from the scientific community, and a panel established by President Kennedy in 1963 supported her claims. *Silent Spring* was also quickly translated into several other languages. Carson thus challenged the dominant thinking at the time that the many new chemicals put on the market since the 1920s were harmless to nonpests and humans. Her book forced many scientists, policy makers, and the public to fundamentally rethink the use and regulation of pesticides.

In addition to its use in agriculture, DDT was extensively applied in indoor house spraying against pests and for coating of bed nets against malaria-carrying mosquitoes. The connection between different kinds of *Anopheles* vectors and malarial transmission had been known since 1897. Early malaria control efforts focused on local environmental management measures that included the elimination of mosquito habitats in wetlands through drainage and landfills, the control of larvae by predatory fish in ponds, and the killing of larvae with oil and the pesticide Paris Green (Mabaso, Sharp, and Lengeler 2004). However, these techniques were largely superseded by chemical controls with the introduction of DDT. In 1955, as hundreds of millions of malaria cases every year led to millions of deaths, the WHO began coordinating the Global Malaria Eradication Programme, which relied heavily on the use of DDT.

Although the Global Malaria Eradication Programme helped eliminate malaria in temperate climate zones, it failed to tackle the malaria problem in large parts of the tropics, in particular in sub-Saharan Africa. Because parasite development is optimal between 25°C and 30°C (and ceases below 16°C), the program was unsuccessful where malaria posed the largest problem. The Global Malaria Eradication Programme was abandoned in 1969 for several reasons, including environmental concern about DDT, growing vector resistance to DDT, and donor fatigue, leading to a sharp decline in funds for malaria abatement (Sachs 2002). Shifting focus from eradication to management, the WHO in 1992 launched the Global Malaria Control Strategy (Mabaso, Sharp, and Lengeler 2004). Building off this, the WHO, UNICEF, the UNDP, and the World Bank in 1998 initiated the Roll Back Malaria Partnership to fight malaria.

In the wake of Rachel Carson's warnings, other researchers began voicing similar concerns about other chemicals. The first scientific warnings about PCB-like substances had been made as early as 1899 as the skin disease chloracne in workers was linked to exposure of chlorinated organic chemicals. Health effects in workers exposed to PCBs in the early 1930s were addressed through the use of protective clothing, which garnered only limited public attention (Koppe and Keys 2002). In the mid-1960s, Sören Jensen expressed concern over PCBs entering the air and causing environmental contamination ("Report of a New Chemical Hazard" 1966). Jensen took samples from wildlife as well as from his own hair and that of his wife and his five-month-old daughter: all contained PCBs. Jensen also tested feathers from eagles preserved at the Swedish National Museum of Natural History, which revealed environmental traces of PCBs back to at least 1944 (Jensen 1972).

In addition, industrial accidents drew attention to the risks of hazardous chemicals. Scientists and public health officials focused on the health risks of PCBs when people in Yusho, Japan, in 1968 were poisoned after consuming rice contaminated with high levels of PCBs that had leaked from a faulty heat exchanger. This short-term, high-level exposure to PCB-contaminated rice led to several adverse human health conditions (Kuratsune et al. 1996). Doctors observed a series of different skin problems, including darkening of the skin, hyperkeratosis, and chloracne. Babies who were born to mothers after the accident were smaller than normal and also suffered from various skin problems. In addition, some evidence showed an increase in cancer deaths among local males in particular. Together, the findings by Jensen and the observed effects of the Yusho disaster prompted a swift increase in research into PCB toxicity in many countries (Shifrin and Toole 1998).

Early attempts to evaluate environmental and human health effects of hazardous chemicals such as PCBs and DDT were plagued with many analytical difficulties. This was the result of multiple factors (Shifrin and Toole 1998, Mellanby 1992). First, scientists had little previous experience with analytical techniques, environmental data, and technical equipment to effectively measure and study environmental and health risks of these substances. Second, many commercial chemicals are not a single substance, but exist in various formulations of commercial mixtures, often with different environmental, physical, chemical, and biological effects. Third, sensitivity to most chemicals has proven to vary with species, gender, and age. The analytical challenge associated with this variability both before and after discharge to the environment is "akin to hitting a moving target" (Shifrin and Toole 1998, 248).

All of these factors made it difficult to conclusively link a specific chemical to a particular observed environmental and human health outcome in cases of long-term, low-dose exposure, the most common kind of chemicals exposure (as opposed to instances of short-term, high-dose exposure as in the case of the Yusho accident). These problems impeded policy making and the adoption of control measures. Nevertheless, many Northern industrialized countries began to develop domestic regulatory procedures to assess the use of a small set of hazardous chemicals in the late 1960s and early 1970s. An important challenge for these early assessments was to design reliable ways to identify the chemicals that posed serious environmental and human health threats for regulation, while allowing societies to continue reaping the benefits from the use of less harmful chemicals. This remains a key issue in current assessment work.

The use of Agent Orange and other herbicides by the U.S. military during the Vietnam War in the 1960s and early 1970s also attracted much scientific and public interest.[2] Many of these herbicide mixtures used dioxin as a critical ingredient. Chemically close to PCBs, dioxin is one of the most toxic substances ever produced.[3] It is estimated that over 70 million liters of herbicides containing 366 kilograms of dioxin were discharged from aircrafts, helicopters, boats, trucks, and hand sprayers as part of the Operation Trail Dust program, designed to defoliate forests and mangroves, clear areas around military installations, and destroy local food supplies (Stellman et al. 2003). This led to widespread contamination, causing significant increases in infant mortality, congenital malformation, miscarriage, premature death, and cancer among the up to 5 million people who were directly sprayed, as well as the military personal involved in the program (Palmer 2007).

The Yusho spill and the use of Agent Orange in Vietnam are only two examples from a much longer list of chemical disasters across the world. For example, a chemicals and plastics company in Love Canal near Niagara Falls in upstate New York dumped large amounts of toxic chemicals, including dioxin, into an old canal bed between the 1930s and the 1950s. In the 1970s, these chemicals began leaking through to the school and housing tract that had been built on the old dump site as the area was turned into a working-class neighborhood. Health effects linked to chemicals exposure in the Love Canal case included high birth defect and miscarriage rates, liver cancer, and high incidence of seizure-inducing nerve disease among small children. Love Canal became an important catalyst for the passing of the U.S. Superfund Act in 1980, which addresses the cleanup of contaminated sites (Layzer 2006, Anastas and Warner 1998).[4]

Other highly publicized industrial accidents involving hazardous chemicals happened in Italy and India in the 1970s and 1980s. In Italy, a chemical factory in the town of Seveso accidentally released large amounts of dioxins into the air in 1976. This release killed many farm animals, and 2,000 people were treated for dioxin poisoning at local hospitals.[5] The major leakage of methyl isocyanate gas from the Union Carbide factory in the Indian city of Bhopal in 1984 caused even more devastation. The release of this highly poisonous gas into the open air resulted in the immediate death of almost 4,000 people who lived next to the factory, and several thousand other local people experienced permanent or partial disabilities.[6] The Love Canal, Seveso, and Bhopal disasters—together with other local accidents all over the world—stimulated many domestic and international efforts to better protect workers and reduce industrial releases.

Growing trade in hazardous chemicals and wastes from industrialized to developing countries furthermore contributed to local management and human health problems in the 1970s. In addition to the regular trade in commercial chemicals, pesticides that were discarded or banned in industrialized countries were regularly sold to or dumped in developing countries. This practice not only exposed handlers in developing countries to great hazards, it also posed a "circle of poison" risk to consumers in industrialized countries. This notion refers to the process by which a hazardous pesticide that has been banned for use in one country is still lawfully produced and exported, and then returns to the exporting country in the form of residues in imported vegetables, fruits, and other food products (Emory 2001, 48). Thus, the "circle of poison" is one way in which chemicals production, use, and exposure in different parts of the world are connected.

Recently, a large number of analysts and policy advocates have expressed great concern about the sharp increase in the transfer of e-wastes from industrialized countries to developing countries for recycling and disposal (Iles 2004). Much of this e-waste contains a multitude of hazardous chemicals and heavy metals that pose significant environmental and human health risks during the entire management chain, from initial export to final disposal. This growing practice of exporting e-waste from high-consumption societies in the North gives rise to important environmental justice and human security questions. Specifically, environmental justice advocates are concerned about the violation of the basic rights of marginalized and poor people in many parts of the global South who live next to enormous dump sites or work in the waste recovery business and who are exposed to high levels of hazardous substances leaking from unsafely stored and handled e-waste (Pellow 2007).

In addition to the production and use of pesticides and industrial chemicals, toxic chemicals can form as by-products of combustion processes, industrial production, and waste management. Some by-products like dioxins and hexachlorobenzene (HCB) have also been intentionally manufactured, but as their direct uses have been phased out, much regulatory focus has shifted to managing them as inadvertent by-products (e.g., most current domestic and international regulation of dioxin focuses on minimizing unintentional releases). In contrast, other major hazardous by-products such as polycyclic aromatic hydrocarbons (PAHs) and furans, have never been produced for direct use. Regulatory efforts on by-products set technical emission standards requiring the use of best available techniques or require the use of alternative production and management processes to minimize emissions.

From Local to Long-Range Exposure

Scientists expressing concerns about chemicals in the 1960s were primarily worried about the effects of local exposure. At the time, environmental and human health problems had been documented in workers and direct handlers; people subjected to short-term, high-dose exposure; and wildlife in geographical areas of direct application of pesticides. This local perspective was the dominating scientific and policy view of the chemicals problem throughout the 1970s. Starting in the 1980s, a growing body of scientific data suggested that emissions of hazardous chemicals could also travel long distances and cause adverse environmental and human health effects far from their origin of release. This realization, which further transformed the chemicals issue into a distinct transboundary issue, stemmed from a series of largely unintentional and unanticipated discoveries (H. Selin 2000, 2003).

Many contamination studies in the 1980s were conducted in the Arctic. During the early phases of the Cold War, the North Atlantic Treaty Organization built military radar stations across northern Canada as part of an early-warning radar system against Soviet missiles that could cross the Arctic region toward North American military and civilian targets. In the late 1970s, these radar stations were becoming obsolete due to advancements in military technology and new defense strategies, and were gradually abandoned. By the mid-1980s, Canadian government officials realized that the deserted radar stations likely contained PCBs, possibly causing local contamination problems. A research program was launched to assess the environmental conditions around the deserted radar stations. To assess PCB levels in soil and fish, measurement samples were taken around the stations.

When conducting this type of study to assess contamination levels, scientists typically use comparative background samples from other areas that are believed to be uncontaminated to compare contamination levels between "pure" and "contaminated" areas. For the radar station study, distant areas in Arctic Canada that were thought to be pristine and free from chemical pollution were chosen as background control sites for comparison. However, the initial analysis surprisingly showed that samples from the remote areas contained high levels of PCBs. At first, the findings were thought to be due to measurement error or laboratory contamination of the samples, but a second testing of the background sites also demonstrated the presence of surprisingly high levels of PCBs (as well as in the areas around the abandoned radar stations that were initially the primary scientific and policy interest).

The unexpected measurement results from the background sites used in this study stimulated increased scientific attention on the issue of long-range transport of chemical emissions to the Arctic. Spurred on by both qualitative and quantitative improvements in data samples and measurement and analytical techniques, three significant factors relating to a set of hazardous persistent organic substances with important scientific and policy implications were revealed in the late 1980s: evidence of systematic long-range atmospheric transport of emissions to the Arctic; discoveries of high environmental contamination levels throughout the Arctic region; and indications of actual and potential environmental and human health implications, particularly among local indigenous populations who were exposed primarily through dietary intake of local wildlife. Although unexpected, this discovery was consistent with earlier studies.

Holden (1970) was among the first to measure organic chemicals in Arctic Canada. In a study on polar bears between 1968 and 1972, Bowes and Jonkel (1975) identified PCBs and DDT in wildlife throughout the Canadian Arctic.[7] Uncertainty existed, however, about emission sources and transportation pattern. Bowes and Jonkel speculated on the importance of several human activities that could cause emissions, such as shipping, exploration, and drilling, and they proposed a multitude of possible carriers of emissions, including oceans, rivers, air currents, and animals. The first suggestion that substances like DDT and PCBs systematically migrate through the atmosphere and condense in low-temperature regions came in 1974 (Wania and Mackay 1996). A decade later, Oehme and Manø (1984) presented "strong proof" of long-range atmospheric transport of organic pollutants to the Arctic. Their findings suggested that pollutants from both North America and more distant regions reached the Arctic.

At the same time as evidence of extensive long-range atmospheric transport of hazardous persistent organic substances became frequent in the 1980s, several studies demonstrated the presence of toxaphene, chlordane, HCH (hexachlorocyclohexane), dieldrin, HCB (hexachlorobenzene), and mirex in wildlife in remote Arctic areas. Many of these compounds are among the first generation of synthesized organic chemicals from the 1940s and 1950s. Subsequent studies confirmed that atmospheric input was the main transport pathway (Muir et al. 1988, 1990). While environmental concentrations of DDT and PCBs seemed to be declining, though they were still present in biota, levels of other chemicals were increasing. Sometimes this was the result of a ban of one hazardous substance leading to the use of another. For example, chlordane, which is now classified as a POP, was used to replace DDT following its ban (UNECE 1994).

Other studies showed that the Arctic situation and its combination of interacting factors relating to long-range transport and environmental exposure were not unique. The Arctic data were supported by information from other geographical areas. For example, studies in the Baltic Sea region in the 1980s indicated that long-range atmospheric transport was a major route by which DDT and PCBs entered sensitive terrestrial and aquatic ecosystems (Larsson and Okla 1989). In addition, Baltic Sea chlordane concentrations were found to have risen sharply, indicating that pollutants previously detected in only very small quantities had increased in relative significance. These and other findings contributed to a growing scientific and political interest in the 1990s in substances that are toxic and persistent in the environment, accumulate in individuals, and magnify, or concentrate, further up food webs.

It was not a coincidence that most early data on long-range transport and environmental accumulation came out of the Northern Hemisphere. Much of the expertise and technology needed to identify and measure this phenomenon was located in North America and Europe. It was also mainly North American and European countries that had developed the most extensive domestic chemicals legislation and devoted resources to assessment and regulation as public apprehension about the widespread use of toxic chemicals grew larger. In addition, global wind current patterns carry emissions from the tropics toward the poles in what has been described as a series of "grasshopper" movements, in which compounds volatilize and condense with seasonal temperature changes at midlatitudes as they move toward the polar regions (Wania and Mackay 1993, 1996). This contributes to the accumulation in the Arctic of hazardous chemicals originating from all over the Northern Hemisphere.

Several of the scientific findings on hazardous chemicals in the 1980s and 1990s also focused on human health risks from continuous exposure. In a risk assessment of hazardous chemicals affecting the Canadian Arctic human population, a study released in 1988 showed that 15.4 percent of the men and 8.8 percent of the women ingested more PCBs than the national conditional tolerable daily intake set by Health and Welfare Canada (Kinloch and Kuhnlein 1988). This discovery coincided with growing scientific evidence of long-range transport and the discovery of a broad range of toxic organic pollutants in much Arctic wildlife. At about the same time, another survey showed that Inuit women in northern Quebec had PCB concentrations in breast milk that were five times higher than in Caucasian women in southern Canada (Dewailly et al. 1989).

The early findings were supported by a follow-up study investigating Inuit babies born in 1989 and 1990. Examining breast milk samples, the study demonstrated that by virtue of their location at the highest trophic level of the Arctic aquatic food webs, Inuit women from northern Quebec had significantly higher PCBs levels than women who had given birth in the southern part of Quebec (Dewailly et al. 1992). Later scientific studies showed a similar situation for other hazardous organic substances such as heptachlor, HCB, dieldrin, DDT, chlordane, and endrin. Concentrations of these chemicals found in Arctic animal and human populations are among the highest measured anywhere in the world (Dewailly et al. 1993, UNECE 1994). The implications of this exposure were unclear, but the high contamination levels caused serious concern among regulators and health officials as the local conditions faced by Arctic indigenous groups became politically sensitive (N. Selin and H. Selin 2008).

Around the same time as concerns of long-range transport were growing in the Northern Hemisphere, developing countries increasingly debated chemicals issues. This was partly the result of people in the South becoming more aware about human health risks from hazardous chemicals in part because of studies done in the North. Debates in developing countries also combined local and international dimensions of the chemicals problem. Many issues related to the use of hazardous pesticides in agriculture and pest control, such as the use of DDT against malaria-carrying mosquitoes. Chemicals risks affected farmers who were spraying crops with pesticides, consumers of food with high levels of pesticide residues, and people using chemicals for health protection. Many chemicals used in developing countries were imported with little knowledge by authorities, and local users often did not take safety measures (Paarlberg 1993).

In 1996, thirty-four years after the publication of *Silent Spring*, the book *Our Stolen Future* presented the "endocrine disruptor hypothesis" (Colborn, Dumanoski, and Myers 1996), which received much attention among both regulators and environmental advocacy groups. This hypothesis, based on data from both wildlife and humans, suggested that certain synthetic chemicals such as PCBs acted like the hormone estrogen, causing reproductive and behavioral abnormalities, and they therefore threatened the ability of animals and humans to reproduce and develop normally. Another prominent study from 1996 linked the consumption of PCB-contaminated fish by mothers in the Great Lakes area with impaired cognitive functioning in their children (Jacobson and Jacobson 1996). These and other findings spurred a growing interest among regulators and the scientific community in the effects of endocrine disrupters in humans (Krimsky 2000).

Since the 1990s, many political and management efforts have aimed to prevent short-term, high-level exposure to hazardous pesticides and industrial chemicals. Continuing scientific studies also explore the possible effects of long-term, low-dose exposure to "cocktails" of multiple persistent, bioaccumulative, and toxic chemicals to which most people are exposed, albeit to varying degrees. Toxicologists remain uncertain about exactly what human health and environmental effects there may be from such exposure, but possible impacts include developmental problems in fetuses and small children. As a result, public health authorities in many countries have issued dietary recommendations, particularly for pregnant women and young children, to limit their intake of certain foods (mainly fish with a relative high fat content) to reduce chemical exposure. The WHO furthermore set guidelines for the maximum recommended content of hazardous substances in drinking water and food.

Early Policy Responses

The global chemicals regime has been developed primarily since the 1960s, but international chemicals policy dates back much further. The first international instrument addressing chemicals may have been the St. Petersburg Declaration from 1868, banning the use of "fulminating or inflammable substances" in military projectiles weighing less than 400 grams (Shaw 1983, 110). However, this declaration is chiefly part of humanitarian law, rather than an effort to address environmental and human health issues. Early such action on hazardous chemicals focused on risks to workers' health. The ILO in 1919, its first year in existence, issued nonbinding recommendations on the risks of lead and white phosphorus to workers. More recent and comprehensive multilateral policy responses to the environmental and human health threats from hazardous chemicals have been developed with considerable difficulty and controversy. Many of the early legal efforts were regional in scope.

The OECD, created by the world's leading industrialized countries in 1960, was one of the first international organizations to address environmental and human health issues relating to chemicals in a comprehensive manner (Alston 1978, Lönngren 1992). In the late 1960s, the OECD recommended that member states reduce "nonessential" uses of hazardous chemicals, exercise more care in the application of persistent pesticides, and develop less harmful pesticides and alternative means of pest control. In 1971, the OECD established its Sector Group on the Unintended Occurrence of Chemicals (renamed the Chemicals Group in 1975) to manage chemicals in ways that protected the environment and human health "while avoiding negative effects for the economy and trade" (OECD 1981, 22). These management efforts included expanded information exchange and the development of common technical standards among member states for testing and evaluation.

The OECD did more than merely engage in cooperation on technical issues; it also worked to control the use of chemicals determined to pose unacceptable environmental and human health risks. As a result, its political efforts in the early 1970s were among the first international regulatory measures on hazardous chemicals. These actions included a decision by the OECD Environment Council in 1973 to restrict the use of PCBs to a few identified areas (OECD 1973a). A report by the OECD Environment Directorate from the same year took a progressive precautionary approach relatively uncommon at the time. The report stated that although all aspects of PCB toxicity were not fully known and that measured environmental quantities did not represent an immediate human health hazard, their persistent nature and the risk of environmental accumulation justified introducing control measures even before full knowledge was available (OECD 1973b).

Several other international organizations and forums that also included developing countries addressed chemicals in the 1970s as well. The UN Conference on the Human Environment in Stockholm in 1972 paid much attention to hazardous substances. The Stockholm Action Plan adopted at the conference stressed the need for intensified cooperation and research on hazardous chemicals, called on states to minimize environmental releases of such substances, and urged governments to promote corporate and public awareness and establish improved testing procedures. Furthermore, the WHO created its Environmental Health Criteria Programme in 1973 to assess environmental and human health data and provide guidelines for exposure limits (Lönngren 1992). In addition, the WHO and FAO worked together in the Codex Alimentarius Commission on recommendations for setting maximum pesticide residues in food (Paarlberg 1993).

UNEP, created in 1972, quickly became a central node for much international legal, political and technical cooperation on hazardous substances (Alston 1978, Lönngren 1992). Based on several recommendations in the Stockholm Action Plan on the importance of developing internationally harmonized procedures for assessing and managing hazardous substances, UNEP created the International Register of Potentially Toxic Chemicals (IRPTC) in the mid-1970s to collect and disseminate information about domestic regulations on chemicals. The Stockholm Action Plan also stressed the need for making more resources available to developing countries for domestic capacity building to better manage chemical risks. The Stockholm Action Plan furthermore stimulated the creation of the International Programme on Chemical Safety (IPCS), and provided motivation for expanding work on hazardous chemicals in the ILO and OECD (Lönngren 1992).

In the 1970s and early 1980s, UNEP facilitated the creation and implementation of several multilateral agreements that sought to protect oceans and regional seas from dumping and other types of pollution, including toxic chemicals (P. Haas 1990, UNEP 2002a). These included some legal efforts that had begun already before UNEP was created, such as the 1972 International Convention on the Prevention of Marine Pollution by Dumping of Wastes and Other Matter (London Convention), and the MARPOL Convention, which includes the 1973 International Convention for the Prevention of Pollution from Ships and its 1978 Protocol. These efforts to develop international marine pollution prevention law took note of language in the Stockholm Declaration and the Stockholm Action Plan on the need for improved protection of the marine environment from a wide range of hazardous substances and other kinds of pollution.

One of UNEP's first major accomplishments was the establishment of the Regional Seas Programme in 1974, which was based on discussions at the Stockholm Conference and several political and technical priorities outlined in the Stockholm Action Plan. Under this program, countries began to draw up and implement protection plans for shared seas, including pollution-preventive measures designed to reduce adverse effects from high aquatic levels of hazardous substances. These activities also included participation by many IGOs and NGOs. Supported by the Regional Seas Programme, collective responses were often formed under the framework of legally binding conventions, sometimes with associated protocols. A Regional Seas Programme Activity Centre was set up in Geneva in 1977 to coordinate the work carried out under the program, which remains in operation.[8]

European states in particular also negotiated several regional sea and river agreements outside the Regional Seas Program in the 1970s. These agreements introduced legally binding restrictions and bans on the use and discharge of hazardous chemicals, including PCBs, DDT, and others exhibiting similar characteristics (Pallemaerts 2003, H. Selin and VanDeveer 2004, Eckley and H. Selin 2004). Such treaties include the 1972 Convention for the Prevention of Marine Pollution by Dumping from Ships and Aircraft (the Oslo Convention) and the 1974 Convention for the Prevention of Marine Pollution from Land-based Sources (the Paris Convention) covering the Northeast Atlantic. In 1974, the Baltic Sea littoral states negotiated the Convention on the Protection of the Marine Environment of the Baltic Sea Area (the Helsinki Convention). The Rhine River countries in 1976 adopted the Convention on the Protection of the Rhine against Chemical Pollution, drawing on collaborative efforts since the 1950s (Mingst 1981).

The EU began to develop chemicals legislation in the 1960s. Its first chemicals directive, from 1967, attempted to harmonize requirements for the classification, packaging, and labeling of dangerous substances across all member states. This was followed in 1976 by the adoption of a directive on the restriction of marketing and use of chemicals, which marked the beginning of more regulatory EU efforts on hazardous chemicals, which have since been significantly expanded (H. Selin 2007). In addition, Canada and the United States signed the Great Lakes Water Quality Agreement in 1972, which was updated in 1978. Under this agreement, implemented under the supervision of the International Joint Commission, Canada and the United States pledged to protect the Great Lakes basin ecosystem through the "virtual elimination" of discharges of all persistent toxic substances (N. Selin and H. Selin 2006).

In the early 1980s, several IGOs, countries, and environmental advocacy groups voiced concerns that the many efforts to date were too scattered and largely inadequate to address the multifaceted problem of hazardous chemicals. This triggered efforts to expand international regulations, including on trade. Leading industrialized countries and the chemicals industry, however, were reluctant to restrict the profitable trade in chemicals. Instead, UNEP and FAO initiated efforts focused on the creation of a mechanism to gather and disperse data on the trade in hazardous chemicals to give countries more information about imported chemicals. These discussions led to the development of two separate mechanisms: the 1985 International Code of Conduct for the Distribution and Use of Pesticides and the 1987 London Guidelines for the Exchange of Information on Chemicals in International Trade (Pallemaerts 1988, 2003; Krueger and Selin 2002).

The Code of Conduct and the London Guidelines, both voluntary, were intended to diffuse risk assessment information to enable countries to better assess chemical hazards. To this end, they incorporated a PIC principle: the country exporting a chemical had to submit a notification prior to export, and the importing country had to express consent prior to import. In addition, the international transport of hazardous wastes gained increasing political attention. As part of this trade, hazardous chemicals that were discarded or banned in industrialized countries were exported to developing countries, which often lacked the capacity to manage them. In response, the UNEP Governing Council in 1987 adopted the voluntary Cairo Guidelines and Principles for the Environmentally Sound Management of Hazardous Wastes, based on a PIC principle similar to the mechanisms for trade in chemicals.

Nevertheless, many developing countries and environmental advocacy groups, supported by a few industrialized countries, argued in the 1980s that these voluntary measures targeting hazardous chemicals and wastes were not strong enough to adequately protect developing countries from hazardous chemicals and wastes (Kummer 1995, Krueger 1999, Brikell 2000). As a result, supporters of expanded management efforts and tougher controls called for the creation of international legal regulations on the transboundary transport and trade in hazardous wastes. In response to these political pressures, UNEP's Governing Council in 1987 authorized the start of treaty negotiations aimed to strengthen the PIC requirements outlined in the Cairo Guidelines (the same year the voluntary guidelines were adopted). These negotiations lead to the adoption of the Basel Convention in 1989, which remains the main international instrument dealing with waste issues (see also chapter 4).

The Intensification of Policy Making

Starting in the early 1990s, policy efforts on hazardous chemicals expanded greatly. Chemicals were a key issue at the UN Conference on Environment and Development (UNCED) in Rio de Janeiro in 1992. Chapter 19 of agenda 21 adopted at UNCED identified six priority areas for chemicals management. Three of these focused on data issues: (1) expand and accelerate international risk assessments, (2) harmonize chemical classification and labeling across countries and registries for easier identification, and (3) stimulate international exchange of information on hazardous chemicals. The other three priority areas aimed at strengthening organizational structures and capabilities: (4) establish more effective risk-reduction programs, (5) strengthen national capacity and capability for better chemicals management, and (6) prevent illegal international traffic in hazardous and dangerous products.

Signaling that improved chemicals management remained a priority, the Commission on Sustainable Development (CSD), created in the aftermath of UNCED to continuously review progress on and guide efforts on promoting sustainable development, made hazardous chemicals and wastes a key issue during its first session in 1994 (DeSombre 2006). Countries also established the IFCS in 1994 in accordance with chapter 19 of agenda 21. The IFCS created a platform for debates between governments, IGOs, industry groups, advocacy groups, and scientific experts to discuss management issues and make policy recommendations. In addition, the IOMC was set up in 1995 to help coordinate the efforts of the many international organizations participating in international chemicals management, including WHO, ILO, UNEP, FAO, UNIDO, UNITAR, and the OECD.

Following UNCED, states, IGOs, and NGOs increasingly focused on life cycle issues. This resulted in the creation of three major treaties that are at the center of the chemicals regime: the 1998 Rotterdam Convention (see chapter 5), the 1998 CLRTAP POPs Protocol (see chapter 6), and the 2001 Stockholm Convention on POPs (see chapter 7). In addition, the Globally Harmonized System for the Classification and Labelling of Chemicals, adopted in 2003, sets common criteria for classifying chemicals and developing compatible labeling and safety data sheets for easier identification of chemicals and risks. Countries that lack domestic systems for hazard classification and labeling are urged to adopt the global criteria. Countries that already have such domestic systems are expected to align them with the global criteria. UNITAR, ILO, and OECD work with developing countries to design and carry out implementation plans.

The world's governments at the WSSD in 2002 restated their commitments to agenda 21 and to improve life cycle management of hazardous chemicals. In addition, governments adopted the goal in the Johannesburg Plan of Implementation that chemicals should be "used and produced in ways that lead to the minimization of significant adverse effects on human health and the environment" no later than 2020 (WSSD 2002. para. 23). To this end, the Johannesburg Plan of Implementation called for the rapid ratification and entry into force of the Rotterdam Convention and the Stockholm Convention, as well as expressed support for the Basel Convention and the Globally Harmonized System for the Classification and Labelling of Chemicals. In addition, the Johannesburg Plan of Implementation supported efforts to develop a strategic approach to international chemicals management.

By the late 1990s, many countries, IGOs, and stakeholders expressed concern that the disparate character of the many legal and political efforts on hazardous chemicals was a major impediment toward fulfilling the goals of agenda 21. In response, the UNEP Governing Council in 2001 called for an assessment of a more concerted global approach to chemicals management, working closely with the IFCS. Based on this assessment, the UNEP Governing Council in 2003 initiated the development of a SAICM, which was adopted in 2006. Several aspects of it were controversial, however. For example, while the EU wanted this strategy to be designed as a legally binding agreement, the United States and many other countries opposed these efforts. As a result, SAICM is structured as a voluntary program guiding activities under various chemicals treaties. In addition, industrialized countries rejected a call by developing countries that SAICM should contain a mandatory commitment by donor countries to provide financial resources for its implementation.

SAICM prioritizes several major life cycle issues: improving domestic enforcement of laws, enhancing coordination among different agencies dealing with chemicals, and including a variety of stakeholders in domestic management and decision-making processes. SAICM furthermore stresses the need for building capacity in developing and transitional economies to manage chemicals safely, in part by improving technical cooperation. At the International Conference on Chemicals Management in 2006, where SAICM was adopted, governments agreed on the Quick Start Programme in support of expanded capacity-building activities in developing countries and countries with economies in transition. Fifteen countries were initially selected for program participation.[9] The Quick Start Programme, however, is only a first, and small, step toward implementing SAICM, which is set to be a complex and resource-intensive process.

Conclusion

As IGOs, states, and NGOs continue to work together under SAICM and on many treaty implementation issues, it is important to recognize the multitude of policy developments on hazardous chemicals that have taken place since the 1960s. To understand how these many policy efforts have evolved, it is necessary to look at how the evolution of scientific understanding has influenced policy making. Furthermore, the structure and content of the chemicals regime have been shaped by the formation of coalitions of regime participants around specific issues, leading to policy expansions. With a growing density of treaties and programs, policy making has also increasingly become characterized by policy diffusion, as institutional linkages influence regime effectiveness and multilevel governance. The next four chapters examine these issues as they focus on the creation and implementation of the Basel Convention, the Rotterdam Convention, the CLRTAP POPs Protocol, and the Stockholm Convention, respectively.

4

The Basel Convention and Hazardous Waste Management

This chapter analyzes chemicals policy and management issues related to hazardous wastes. Hazardous chemicals are released into the environment when hazardous wastes are inappropriately stored, transported, recycled, or destroyed. The hazardous waste issue has been on the international agenda for decades, accompanied by a string of high-profile cases of illegal dumping. In one of the most notorious examples, the cargo ship *Khian Sea* went to sea in 1986 in search of a disposal site for 14,000 metric tons of toxic incinerator ash (labeled as fertilizers), which originated from Philadelphia and contained high levels of lead and cadmium. The ship spent almost two years at sea in search of a place to discard its cargo. It sailed to five continents and changed its name twice, before dumping 4,000 tons of ash on a beach in Haiti and the remaining 10,000 tons at sea somewhere between the Suez Canal and Singapore (Krueger 1999, Jaffe 1988, Millman 1988).

Many more current examples demonstrate the multitude of legal, political, and practical problems regarding hazardous waste transfers. For example, during the night of August 19, 2006, the Greek-owned and Panama-registered vessel *Probo Koala*, chartered by the Dutch-based company Trafigura, dumped near Abidjan 500 tons of "a fuming mix of petrochemicals and caustic soda" that originated from the Mediterranean region (Africa Research Bulletin: Economic, Financial and Technical Series 2006; Environment News Service 2006). This illegal dumping caused several deaths and severe health effects in tens of thousands of nearby people. In response to this disaster, local civil servants were fired, and the Dutch firm agreed to pay $200 million in compensation to the Ivorian government amid allegations that the dumped waste was not cleaned up quickly enough, continuing to expose local residents to great risk (Bryant 2007).

Specifically, this chapter examines the creation and implementation of the Basel Convention, which sets out to minimize the generation of hazardous wastes and to control and reduce their transboundary movement. The treaty targets the transport and disposal of chemicals when they fall under the treaty definition of a hazardous waste, or when they are present in other kinds of articles that meet this definition. The chapter highlights several issues associated with the formation of actor coalitions and policy diffusion that are of importance to the effectiveness of the Basel Convention and the chemicals regime: the incorporation of the PIC principle for managing trade and subsequent efforts to strengthen controls; the development of technical guidelines for waste management; the establishment and operation of regional centers supporting implementation and capacity building; and the creation of mechanisms for liability, monitoring, and compliance.

The chapter begins with an analysis of global wastes issues and related human health and environmental concerns, followed by a discussion of the development of early, voluntary policy responses to managing the international trade in hazardous wastes and how these initiatives were linked with efforts to regulate the trade in hazardous chemicals. Next, the chapter examines actor coalitions and policy issues during the negotiations of the Basel Convention, which institutionalized a legal framework for addressing the generation and management of hazardous wastes. This is continued by an examination of the implementation of the Basel Convention, including how governance and actor linkages with other policy processes shape activities under the convention. The chapter ends with a few remarks on the long-term effectiveness of the Basel Convention and its role in multilevel governance under the chemicals regime.

Global Waste Issues

The generation of hazardous wastes has increased sharply since the 1970s, and it continues to grow. Estimates of actual waste levels differ extensively, however. This variation in estimates stems in part from the absence of a globally agreed-on definition of what is considered a waste, as well as what constitutes a hazardous waste (Kummer 1995). There are also large data gaps about the generation of hazardous wastes in many countries (UNEP 2002b). UNEP estimates an annual global generation of 150 million metric tons (UNEP 2002c), but some experts put the figure as high as 500 million metric tons (Kummer 1995).

Generally a waste is an unwanted by-product of industrial and household activity (O'Neill 2000). Under the Basel Convention, "wastes are substances or objects which are disposed or are intended to be disposed or are required to be disposed of by the provisions of national laws" (article 2). Main categories of hazardous wastes include industrial wastes, agrochemical wastes, medical wastes, and household wastes. Yet there is no simple, objective way to define waste. Any definition is inevitably shaped by political, economic, social, and cultural factors: what one society or person regards as a waste can be seen as a resource somewhere else. For example, changing production standards in one country may result in the export of outdated technology from that country to another in which that technology is still permitted. Similarly, domestic legislations in various countries contain different classifications of what constitutes a hazard (Forester and Skinner 1987, Asante-Duah and Nagy 1998).

Industrialized countries generate the vast majority of hazardous wastes. The OECD (2001) estimated that its twenty-five member countries generated 110 million metric tons in the mid-1990s, with continuing increases. Furthermore, rapid economic development and growing consumption in Asia and Latin America are resulting in significant increases in hazardous waste generation (Yang 2008). The links between chemicals management and waste management are strong. Hazardous chemicals often become hazardous wastes, as outdated or used chemicals are treated as waste products and designated for disposal. In addition, hazardous chemicals exist in a wide range of articles: large ships, industrial apparatus, medical equipment, and electronic home goods, for example. Chemicals in such articles are frequently released into the environment during the recycling and disposal of these articles.

Most early international policy actions on hazardous wastes were taken in response to concerns over marine pollution from ships and marine dumping of hazardous materials. The 1954 International Convention for the Prevention of Pollution of the Sea by Oil, which built on cooperative efforts that began as early as the 1920s, established specific marine zones within which ships could not discharge used oil (Caldwell 1996).

This treaty was replaced by the International Convention for the Prevention of Pollution from Ships (MARPOL) from 1973 and its 1978 protocol, which address issues of marine pollution from ships broadly (not just oil pollution) (Caldwell 1996). Several regional agreements on oceans, regional seas, and rivers also ban or regulate the dumping of hazardous substances and wastes, as discussed in chapter 3.

Countries have furthermore developed controls on the dumping of radioactive waste at sea (Ringius 2001). Between 1946 and 1972, a number of states (including the United Kingdom, the United States, the Netherlands, and Japan) regularly disposed of low-level radioactive wastes in oceans as a way of avoiding having to dispose of it domestically. This issue was debated at the 1958 United Nations Conference on the Law of the Sea, but it was not until the 1972 Convention on the Prevention of Marine Pollution by Dumping Wastes and Other Matter (the London Convention) that countries established an international agreement controlling the disposal of radioactive wastes at sea. The London Convention banned ocean dumping of high-level radioactive wastes, but allowed the continued disposal of medium-level and low-level radioactive wastes. These regulations were extended in 1993 with the adoption of a complete ban on marine radioactive waste disposal.

The international trade in hazardous wastes gives rise to important justice issues (Pellow 2007). The first officially recognized transboundary shipment of hazardous wastes took place in the 1970s, but undocumented shipments across national borders likely occurred much earlier. Although most trade in hazardous wastes is among industrialized countries (O'Neill 2000), the North-South trade has attracted the most political attention. The generation of hazardous wastes in industrialized countries was reaching record levels while domestic regulations became more stringent, forcing many older waste disposal facilities to close even as local opposition prevented the building of new ones. As a result, waste management costs grew rapidly in industrialized countries: one study from the late 1980s estimated the average disposal cost for hazardous waste in industrialized countries between $100 and $2,000 per ton, with corresponding costs in Africa between $2.50 and $50.00 (Kummer 1995).

The difference between exporting hazardous wastes and used goods is not clear-cut, which creates important enforcement problems. Not all international transfers of what may be considered a hazardous waste in the

North take place for disposal purposes. Items that are considered wastes in one country are sometimes exported for continued use in another country. For example, old industrial equipment and scrap metal may still have a commercial value elsewhere or may be treated as secondary raw material to be recycled and reused. It is often developing countries that seek foreign used technology and have a larger market for recycled material. In addition to the export of used items from one country to another, in some cases, entire polluting and waste-producing industries have been dismantled and moved from industrialized countries to developing countries (Clapp 2001).

Hazardous chemicals have always been part of the hazardous waste issue, but rapidly growing quantities of e-waste in industrialized and industrializing countries create even closer linkages between chemicals and waste management (H. Selin and VanDeveer 2006). Industrial machines, computers, and home electronics, among other electronic goods, often contain hazardous substances such as lead, chromium, and brominated flame retardants that may be released during recycling and disposal. High-consumption societies are replacing many of these goods with newer models at greater frequency, especially consumer goods, while the old ones are thrown away even though they still work. This has created growing disposal problems in many countries. Some industrialized countries have responded to this problem by exporting e-wastes to developing countries, particularly in Asia (Iles 2004).

Like most other trade in wastes between industrialized and developing countries, the growing trade in e-wastes is driven by economic factors. Large differences in domestic disposal costs, combined with the exploitation of weaker environmental and human health regulations and enforcement in developing countries, have created a lucrative international market for e-wastes. Often, however, environmental and human costs are ignored. In the words of one American entrepreneur in the e-waste trade; "I could care less where they go. My job is to make money" (Goodman 2003). Conditions around many disposal sites are abysmal. As a policy advocate noted in a *60 Minutes* story on e-waste trade from the United States to China where inadequate handling of e-wastes pose significant local problems: "It's a hell of a choice between poverty and poison, and we should never make people make that choice."[1]

The Basel Convention and the Management of Hazardous Wastes

The Basel Convention, adopted in 1989, is the oldest of the four main multilateral agreements under the chemicals regime. Table 4.1 lists important milestones in the creation and implementation of the Basel Convention and associated policy developments. The Basel Convention, which was prompted by concerns about the increase in transnational transport and dumping of hazardous wastes in developing countries, grew out of several concurrent policy developments beginning in the 1970s. Many of these policy initiatives were furthermore closely linked with political efforts to regulate the trade in hazardous chemicals (see chapter 5).

Table 4.1
Chronology of important events in the creation and implementation of the Basel Convention.

Time	Event
June 1987	UNEP Governing Council adopts the Cairo Guidelines
October 1987	First negotiating session of the Basel Convention
March 1989	Basel Convention is adopted
May 1992	Basel Convention enters into force
December 1992	COP1 requests industrialized countries to refrain from exporting hazardous wastes to developing countries for disposal (decision I/22)
March 1994	COP2 adopts a ban on the export of hazardous wastes from OECD countries to non-OECD countries for disposal and recycling (decision II/12)
September 1995	COP3 adopts the Ban Amendment
December 1999	COP5 adopts the Protocol on Liability and Compensation
December 2002	COP6 adopts a ten-year strategic plan COP6 adopts a formal compliance mechanism COP6 adopts the framework agreement on the Regional Centers for Training and Technology Transfer COP-6 adopts the Basel Convention Partnership Programme
July 2009	171 states and the EU have ratified the Basel Convention 64 countries and the EU have ratified the Ban Amendment 9 countries have ratified the Protocol on Liability and Compensation

Pre-Basel Policy Developments

The rapidly increasing generation of hazardous wastes in industrialized countries and the growing waste trade among these countries prompted the OECD Council in 1976 to issue a recommendation on the development of comprehensive domestic waste management policies in all member countries (OECD 1985, Kummer 1995, Brikell 2000). OECD work in the early 1980s led to the development of guidelines for managing transnational movements of hazardous wastes within the OECD area, including a PIC mechanism that was also applicable to transactions with non-OECD countries. Taking on an early intellectual and material leadership role, the OECD in the mid-1980s began preparing a legally binding agreement on the control of transnational movements of hazardous wastes among OECD states. This work was suspended in early 1989 as the Basel Convention was concluded, but policy ideas from the OECD initiative were diffused by member states into the drafting of the Basel Convention text, linking the two policy efforts.

At the same time as the OECD was expanding its activities, UNEP coordinated work on hazardous waste management (Kummer 1995, Krueger 1999, Brikell 2000). In doing so, UNEP, which was often sympathetic to the interests and demands of developing countries, also demonstrated important intellectual and material leadership for setting global standards. In 1982, UNEP's Montevideo Programme for the Development and Periodic Review of Environmental Law identified the handling and disposal of hazardous wastes and related environmental and human health risks as issues requiring increased global political and legal attention. Based on the work by the Montevideo Programme, the UNEP Governing Council in 1982 created a working group to develop new technical guidelines and policy recommendations on the improved management of hazardous wastes for the purpose of better human health and environmental protection.

As a result of the work carried out by the working group, the UNEP Governing Council in 1987 adopted the first global standards on the transnational transport of hazardous wastes, the Cairo Guidelines and Principles for the Environmentally Sound Management of Hazardous Wastes. Among other things, the Cairo Guidelines, which were voluntary, introduced a PIC scheme for all transnational transport of hazardous wastes, similar to OECD regulations on trade in hazardous wastes. A small but growing coalition of mainly developing countries and environmental advocacy groups, however, did not think that the guidelines were stringent enough and pushed for the creation of legally binding regulations. The Organization of African Unity and Greenpeace in particular went so far as to advocate a complete trade ban as a way to protect developing countries, a proposal that gained some support from the Nordic countries.

The coalition supporting more rigorous trade regulations won a small victory when the UNEP Governing Council in 1987, based on a joint proposal by Switzerland and Hungary, decided to convene negotiations on a legally binding agreement to strengthen the requirements outlined in the Cairo Guidelines (Brikell 2000). The treaty negotiations were closely linked with earlier work on the Cairo Guidelines, the voluntary PIC mechanism and the draft legal agreement under development in the OECD, and the establishment of a PIC procedure on the trade in chemicals. Major principles and policy ideas were diffused across policy forums. Nevertheless, groups of industrialized and developing countries, industry organizations, and environmental advocacy groups expressed diverging opinions on how to regulate the trade and disposal of hazardous wastes. As a result, the design of legally binding controls was highly controversial during the treaty negotiations.

Negotiating the Basel Convention

North-South waste trade, one of the key reasons behind developing a legally binding agreement, quickly became a central issue during the negotiations and split countries and stakeholders into two opposing coalitions.[2] Many developing countries, primarily members of the Organization of African Unity, and Greenpeace advocated a complete ban on transferring hazardous wastes across all national boundaries. This bloc of pro-ban members who took on an important intellectual leadership role alongside UNEP argued that only a complete ban would stop the North's "toxic imperialism." They saw the seemingly endless journey of the ship *Khian Sea* together with other similar cases as a clear sign that industrialized countries refused to accept responsibility and could not be trusted in the absence of a trade ban. The idea of a ban also received support from a few northern European countries.

In contrast, most industrialized countries preferred a PIC procedure similar to the one included in the Cairo Guidelines and discussed in the OECD. The world's largest exporters of hazardous wastes at the time were West Germany, Belgium, the Netherlands, Switzerland, and the United States (Kempel 1993). Led by the United States, the anti-ban coalition believed that continued waste trade was economically beneficial and stressed that shipping wastes to other countries sometimes allowed for their environmentally sound disposal at a lower cost. They also argued that a trade ban would likely increase illegal trade. At the same time, many industrialized countries faced contradictory public pressures. While local communities objected to building new waste disposal facilities (the not-in-my-backyard (NIMBY) syndrome), which in effect increased the need for exporting wastes, public opinion was frequently against trade in hazardous wastes.

The pro-trade coalition furthermore included several developing countries that did not join the African countries and the other members of the pro-ban coalition in their calls for a trade ban, but supported the continuing trade in hazardous wastes. The developing countries that were part of the pro-trade coalition considered the waste trade an important dimension of their domestic efforts to promote economic development and industrialization: they believed that waste imports would bring in income as well as opportunities to gain access to material and equipment discarded in Northern countries that still had value in developing countries. UNEP, under the leadership of Mostafa Tolba, who chaired the negotiations for the Basel Convention, also sided with the pro-trade coalition. UNEP openly expressed support for continued but regulated trade in hazardous wastes (Tolba 1990).

Despite arguing that a trade ban was morally right, the pro-ban coalition had to settle for a compromise that made the voluntary PIC procedure mandatory. These policy developments are an early example of how prominent governance and actor linkages may shape policy outcomes. The voluntary PIC principle, which later became mandatory, emerged as the acceptable compromise between coalitions of actors for and against trade bans. Furthermore, actor coalitions linked formally separate policy developments on wastes and chemicals (see chapter 5). Those opposing trade bans did so across OECD and UNEP policy forums, knowing that any concessions they made in one forum would likely affect policy outcomes in other trade-related forums. Similarly, those pushing for trade bans did this simultaneously on waste and chemicals issues, hoping to break the anti-ban coalition by exerting political pressure in multiple forums.

Based on the compromise between the two main actor coalitions, the Basel Convention creates a mandatory PIC scheme where the transport of discarded chemicals is covered if they meet treaty definitions of hazardous wastes; the Basel Convention classifies wastes as hazardous if they belong to certain categories (annex I) and contain certain characteristics (annex III) (see table 4.2). The treaty regulates waste trades for parties with both other parties and nonparties. A party cannot export hazardous wastes to another party without first receiving the explicit consent of the importing state to proceed with the transfer. Waste exports to nonparties are prohibited unless they are subject to an agreement between the exporter and importer that is at least as stringent as the requirements under the Basel

Table 4.2
Examples of materials that may be considered hazardous waste, adapted from Basel Convention annex I and annex III.

Waste streams that can produce hazardous waste	Waste constituents that may be hazardous	Some characteristics of hazardous waste
• Residues arising from industrial waste disposal operations • Tarry residues arising from refining • Chemical substances arising from research and development • Substances and articles containing PCBs • Pharmaceutical products • Wood-preserving chemicals • Organic solvents • Inks, dyes, pigments, paints, lacquers, varnish • Mineral waste oils, emulsions	• Copper compounds • Zinc compounds • Arsenic and arsenic compounds • Metal carbonyls • Mercury and mercury compounds • Lead and lead compounds • Inorganic cyanides • Asbestos • Ethers • Halogenated organic solvents • Acidic solutions or acids in solid form • Cadmium and cadmium compounds	• Explosive • Flammable • Poisonous • Infectious • Corrosive • Toxic (delayed or chronic) • Eco-toxic

Source: Krueger and Selin (2002, 334)

Convention. Exports of hazardous wastes are furthermore prohibited to Antarctica and to parties that have taken domestic legal measures to ban such imports.

Not surprisingly, many advocacy groups and country members of the pro-ban coalition immediately criticized the Basel Convention as too weak in several aspects (O'Neill 2000). First, those who argued for a complete ban on waste transfers from industrialized countries to developing countries during the negotiations believed that the absence of such a ban was a serious shortcoming. Second, many of the same coalition members thought that the treaty's wording on recycling and raw materials created major loopholes for continued uncontrolled trade in hazardous wastes. Third, members of the pro-ban coalition insisted that the Basel Convention did not contain enough provisions for the environmentally safe disposal of wastes. Finally, critics pointed out that the convention did not contain any provisions for assigning liability or ensuring compensation for any environmental damage resulting from the waste trade.

In response to the lack of a comprehensive North-South trade ban under the Basel Convention, African countries negotiated the Convention on the Ban of the Import into Africa and the Control of Transboundary Movements and Management of Hazardous Wastes within Africa, the so-called Bamako Convention (Donald 1992, Kummer 1995, Brikell 2000). The Bamako Convention, adopted in 1991, seeks to prevent dumping of hazardous wastes in Africa by banning the import of hazardous wastes from any outside country. In addition, substances are subject to regulation if they have been declared hazardous and banned or if registration has been refused or cancelled by the country of manufacture or the country of import and transit. The Bamako Convention also calls for reducing the generation of hazardous wastes to a minimum in terms of quantity or hazard potential and to ensure proper domestic disposal of hazardous waste.

Similar regional actions were also initiated outside Africa (Kummer 1995, Brikell 2000). For example, the 1991 Lomé IV Convention bans the trade in hazardous wastes between members of the EU and former colonies in Asia, the Caribbean, and the Pacific. The 1996 Protocol on the Transboundary Transport of Hazardous Wastes under the Barcelona Convention requires countries around the Mediterranean Sea to take "all appropriate measures to prevent, abate and eliminate" pollution that can be caused by transboundary movements and disposal of hazardous wastes. Parties also have an obligation to "reduce to a minimum" and "where possible eliminate" the transboundary movement of hazardous wastes in the Mediterranean Sea. The 1995 Waigani Convention bans the import of hazardous and radioactive wastes to the island countries in the South Pacific region. Some regional free trade agreements also contain provisions on hazardous waste transfers and management.

Implementing the Basel Convention

When the Basel Convention was adopted, its trade compromise seemed to satisfy very few countries: only thirty-three countries signed the Basel Convention at the diplomatic conference in 1989. All African countries present at the meeting demonstrated their displeasure by refusing to sign. The main reason that the convention entered into force in 1992, three years after its creation, was that it required merely twenty ratifications. The number of parties to the convention, however, has increased sharply over the years, and 171 countries and the EU were parties to the Basel Convention as of late 2009. The United States is the only major industrialized country that is not a party. The United States is also a leading exporter of hazardous wastes, including e-waste.

The secretariat of the Basel Convention is located within UNEP's offices in Geneva. Nine COPs were held between 1992 and 2008.[3] Since its entry into force, the parties have carried out many important legal and organizational developments, several of which were possible only after lengthy and difficult negotiations. Some major issues are furthermore subject to ongoing debate as coalition politics continue. Five sets of interrelated implementation issues that the parties have addressed during the first two decades of the Basel Convention are expanding trade regulations, establishing liability, enhancing compliance, improving management, and establishing the Basel Convention regional centers for capacity building. Many of these issues also connect through governance and actor linkages to other policy-making and management efforts under the chemicals regime as they shape the effectiveness of much multilevel governance.

Expanding Trade Regulations

The strong criticism from the pro-ban coalition against the Basel Convention heavily influenced subsequent policy making during the first few COPs. At COP1, in 1992, most African countries and the Nordic countries, together with Greenpeace and other environmental advocacy groups, continued playing an important intellectual leadership role as they reiterated many of their earlier demands from the convention negotiations in support of (at least) a North-South trade ban (Kummer 1995, Brikell 2000). Although the pro-ban coalition was still unable to convince other parties to accept a trade ban, it successfully pushed for the adoption of decision I/22. This decision, a small step forward for the pro-ban coalition, requested that industrialized countries refrain from exporting hazardous wastes to developing countries for final disposal.

Decision I/22 also secured the continued inclusion of the North-South trade issue on the convention agenda. At COP2, in 1994, the pro-ban coalition continued campaigning for a mandatory ban on exports to developing countries (Kummer 1995, Brikell 2000). Building on the small victory at COP1, the Group of 77 and the Nordic states spearheaded a proposal that resulted in a decision to immediately ban the export of all hazardous wastes from OECD countries to non-OECD countries for final disposal and to prohibit by the end of 1997 the export of hazardous wastes intended for recycling (decision II/12). However, decision II/12 was not formally incorporated into the Basel Convention, and some members of the pro-ban continued to press for stronger legal requirements under the treaty.

Encouraged by minor but nevertheless noteworthy achievements at the first two COPs, the pro-ban bloc at COP3 in 1995 reiterated their support for a North-South trade ban (Krueger 1999, Brikell 2000). This time, the pro-ban coalition was more successful. Responding to growing criticism that decision II/12 was not stringent enough, parties at COP3 adopted the Basel Convention Ban Amendment despite strong opposition from, among others, the United States, South Korea, Australia, Canada, and major industry organizations. The Ban Amendment, which was formally incorporated into the Basel Convention, prohibits the export of hazardous wastes for final disposal and recycling from OECD countries, EU members, and Liechtenstein (listed in annex VII) to all other parties (i.e., developing countries). The annex VII countries, however, remain free to continue trading among themselves under the Ban Amendment.

The main reason for using annexes in the Ban Amendment for dividing parties into two trade-related groups—rather than the earlier OECD/non-OECD categories of Decision II/12—was so non-OECD countries could retain the option of receiving hazardous wastes from industrialized countries by joining annex VII; it is easier for a non-OECD country to move to annex VII under the Basel Convention than to join the OECD. Parties also agreed that there would be no changes to the membership of annex VII until the Ban Amendment entered into force. However, because of an apparent desire by many industrialized countries and developing countries to maintain the economically valuable trade in hazardous wastes for recycling and disposal, ratification of the Ban Amendment has been slow and has not yet become legally binding.

In addition, legal and political uncertainties linger regarding the interpretation of article 17(5) of the Basel Convention concerning the requirement for entry into force of the Ban Amendment (Earth Negotiations Bulletin 2008a). Article 17(5) states that any amendment to the Basel Convention shall enter into force when "at least three-fourths" of the parties have ratified. By late 2009, 65 parties had ratified the Ban Amendment. Basel parties, however, are divided over how article 17(5) should be interpreted, with opponents of the Ban Amendment acting to delay its entry into force. One group of parties argues for a "current time" interpretation: that the three-fourths threshold should be calculated based on the latest number of convention parties (172 by late 2009, with the likely possibility that the number of parties increases in the future). If this position is followed, the Ban Amendment will not enter into force for a long time (if ever).

A second group of parties advocates for a fixed-time approach, believing that the three-fourths threshold should be calculated based on the number of Basel Convention parties when the Ban Amendment was adopted (eighty-two parties by COP3). This group is further divided into two subcamps. The first subcamp believes that the criteria for entry into force have already been met since sixty-five out of eighty-two is more than three-fourths (sixty-two are needed). However, many of those that have ratified the Ban Amendment were not among the eighty-two parties by COP3 but joined the convention after 1995. So a second subcamp argues that it has to be three-fourths of the eighty-two parties by COP3 that have to ratify the Ban Amendment. Among those that were parties by COP3, only 44 had ratified by COP9. Thus, based on this interpretation the number of Ban Amendment ratifications still falls well short of the sixty-two needed.

As the battle over the Ban Amendment continues, major environmental NGOs such as Greenpeace and the Basel Action Network are active coalition partners alongside many developing countries and the EU. In its support for the Ban Amendment, the Basel Action Network hosts a Ban Amendment "Hall of Shame" on its Web site seeking to draw negative attention to countries and organizations that the organization believes are actively working to undermine the Basel Amendment.[4] In addition to the United States, which is not a Basel party, noted members of the Hall of Shame included Basel Convention parties like Australia, Canada, and New Zealand, as well as major industry organizations such as the International Chamber of Commerce and the International Council on Metals and Mining. In contrast, the EU adopted regional regulations implementing the Ban Amendment already in the 1990s.

Establishing Liability

Developing country representatives have long complained that continuing shipping of hazardous wastes is made worse by the unwillingness of industrial countries and firms to repatriate illegally dumped wastes and pay for cleanup and handling. In response, parties at COP5 in 1999 adopted the Protocol on Liability and Compensation, which mainly addresses concerns by developing countries that they often lack sufficient funds and technologies for coping with consequences of illegal dumping or accidental spills (Brikell 2000, Earth Negotiations Bulletin 1999a). The main objective of this protocol "is to provide for a comprehensive regime for liability and for adequate and prompt compensation for damage resulting from the transboundary movement of hazardous wastes and other wastes and their disposal including illegal traffic in those wastes" (article 1).

The Protocol on Liability and Compensation identifies who is financially responsible in the event of an incident during the transport of any kind of hazardous waste covered by the Basel Convention. Each phase of often long and complicated transport chains—from the point at which the wastes are originally loaded on their first means of transport through their domestic export, international transit, country import, up until their final disposal—is covered by the agreement. The Protocol on Liability and Compensation, however, has not yet entered into force, even though it was adopted more than a decade ago and requires only twenty ratifications: there were only nine parties by late 2009. This sends a clear signal that many industrialized and developing country parties are reluctant to accept formal liability for environmental and human health damages resulting from the international trade in hazardous wastes.

Thus, once again, a coalition led by developing countries ensured the adoption of a major policy but is failing to secure its entry into force. Compared to the Ban Amendment, however, there is more widespread opposition also from developing countries engaged in the waste trade to the Protocol on Liability and Compensation. In addition, the politics of the protocol is influenced by parties' concerns about how expanded liability under the Basel Convention may shape similar issues under other treaties. If parties accept strong liability commitments under the Basel Convention, this may shape international law and raise demands for more far-reaching liability mandates in other policy areas within and outside the chemicals regime, linking these issues across policy forums. Many countries remain hesitant to go down that path, which affects their willingness to ratify the Protocol on Liability and Compensation.

Enhancing Compliance
Related to the effort to expand regulations and liability, the Basel parties in 2002 established a compliance mechanism. Because many countries continue to resist interference in domestic waste management practices and international transfers, the creation of the compliance mechanism was possible only after protracted negotiations over the course of several COPs. The compliance mechanism is also designed as a rather weak non-binding tool lacking independent enforcement capabilities. It is therefore intended only to create a more permanent structure for gathering data and monitoring the generation and transnational transport of hazardous wastes. To these ends, the compliance mechanism attempts to standardize reporting across countries and improve reporting from member states—two measures that were deemed necessary after highly scattered

and incomplete national reporting during the first decade of the Basel Convention.

A compliance committee working with the Basel secretariat administers the compliance mechanism. The fifteen committee members are geographically elected, with three members each from the five traditional UN regions: Africa, Asia, Latin America and the Caribbean, Western Europe and others, and Eastern Europe. Many parties protective of their sovereignty were reluctant to accept anything but self submissions. However, in the end, the decision was that submissions to the compliance committee can be made by a party regarding itself, a party regarding another party, or the secretariat. The compliance committee, which meets behind closed doors, provides advice on domestic legal, political, administrative, and technical measures to improve compliance. The committee may also review general issues of compliance and implementation under the Basel Convention. In addition, it may recommend that the COPs act on particular issues to improve compliance (Earth Negotiations Bulletin 2002a).

The overall effectiveness of the compliance mechanism remains largely unproven, in part because of its rather weak nature and the fact that the compliance committee meets in private and has little external transparency. Similar to the political situation surrounding the ratification of the Protocol on Liability and Compensation, the parties' decision to establish only a nonbinding mechanism was shaped by their unwillingness to set a strong precedent for how compliance issues may be addressed under the major chemicals treaties, as well as in other environmental issue areas. Compliance issues have also become central in the implementation of other chemicals treaties as debates in external forums are shaped by decisions on compliance issues under the Basel Convention (see chapters 5 and 7). In this respect, the creation, operation, and effectiveness of compliance mechanisms under several treaties are shaped by a multitude of governance and actor linkages.

Improving Management

Over the years, the parties have worked to clarify which kinds of wastes are and are not covered under the Basel Convention by listing waste categories that fulfill treaty definitions of "wastes" and "hazardous," and developing technical guidelines for managing priority waste streams (Krueger 1999; Brikell 2000; Earth Negotiations Bulletin 2004a, 2006a, 2008a).

The parties at COP4 succeeded in identifying priority waste streams in two new annexes, further specifying the scope and priories of the Basel Convention. Annex VIII lists specific wastes characterized as hazardous under the Basel Convention, and annex IX lists wastes not covered by the treaty. A third list, which is not kept in an annex but is maintained by the Basel secretariat as a working list, contains wastes awaiting classification and possible listing under either annex VIII or annex IX.

The establishment of technical guidelines on the management of chemicals wastes is closely related to other expanding regulatory and management efforts on chemicals wastes in multiple policy forums. Since 2002, work on developing technical guidelines has been guided by a strategic plan setting a decade-long agenda for implementing the Basel Convention. Among other things, the plan stresses the need to tackle priority wastes, including hazardous chemicals and goods containing toxic substances (Earth Negotiations Bulletin 2002a, 2004a, 2008a). Basel Convention technical guidelines on POP wastes have an impact on management efforts under the Stockholm Convention and the CLRTAP POPs Protocol. Conversely, regulations under the two POPs treaties and the Rotterdam Convention influence work on waste streams under the Basel Convention; clearly the effectiveness of all these efforts is tied closely together.

The environmentally sound management of e-wastes is attracting growing attention under the Basel Convention and in many countries all over the world. UNEP (2005) estimates that somewhere between 20 and 50 million tones of e-wastes are produced annually worldwide, and the problem of "digital dumping" of e-wastes from industrialized countries in developing countries is growing. At COP8 in 2006 and COP9 in 2008, many parties stressed that better management of e-wastes was imperative and needed to be considered in connection with the implementation of the strategic plan (Earth Negotiations Bulletin 2006a, 2008a). This heightened focus on e-wastes increases linkages between the Basel Convention and other parts of the chemicals regime that address the use of and trade in industrial chemicals and electronic goods, shaping decision making and implementation across multiple policy forums.

Basel Convention efforts to improve e-waste management involve the promotion and development of voluntary public-private partnerships between governments, industry, and environmental NGOs since COP5 in 1999. These partnerships are intended to aid the environmentally sound management of particular types of wastes. The most extensive such partnership to date is the mobile phone partnership initiative from COP6 under the Basel Convention Partnership Programme (Earth Negotiations Bulletin 2002a, UNEP 2002c). This partnership, which serves as test case for others, seeks expanded participation and responsibility by major industry players such as Motorola, Nokia, Philips, Samsung, and Sony Ericsson in the end-of-life management of mobile phones. This partnership began with the launching of a series of small field projects in Egypt, Senegal, and Vietnam.

In addition, parties are focusing on waste minimization issues. Although waste minimization is a stated goal of the Basel Convention, global waste generation has continued to rise since the treaty was adopted. In response, parties since the early 2000s have worked to strengthen a life cycle approach to waste management toward the goal of achieving waste reduction, which also connects with expanded life cycle management on hazardous chemicals in other policy forums. This includes greater use and dissemination of cleaner production methods and increasing the responsibility of firms to reduce waste levels. Yet the parties have not set any quantifiable targets for the reduction of wastes on any specific categories, including e-wastes. Instead, the COPs have merely debated these issues and endorsed the development of voluntary public-private partnerships designed to expand producer responsibility (Earth Negotiations Bulletin 2006a).

Establishing the Basel Convention Regional Centers for Capacity Building

Closely connected to efforts to improve management, the secretariat, the parties, and other stakeholders have collaborated to create a group of Basel Convention regional centers since the early 1990s. The centers were established under a framework agreement adopted at COP6 in 2002 based on article 14 of the Basel Convention, which calls for the creation of regional centers "for training and technology transfers regarding the management of hazardous wastes and other wastes and the minimization of their generation." The centers are intended to work on a range of capacity-building issues. By 2008, fourteen regional centers had been established in different parts of Latin and South America, Africa, Asia, and

Europe.[5] Additional centers may be created in the future, including one in South Asia (Earth Negotiations Bulletin 2008a).

The regional centers are designed to work in conjunction with the compliance mechanisms and toward the fulfillment of goals identified in the strategic plan. To these ends, they facilitate implementation of the Basel Convention in developing countries and countries with economies in transition by improving regional and domestic management capacities. More specifically, they focus on issues such as promoting environmentally sound waste management, training customs officials and other local people working on waste-related issues in identifying and handling hazardous wastes, facilitating the diffusion of cleaner production technologies to member countries, aiding in national data collection and reporting to the secretariat, and engaging in public education and awareness raising (Basel Convention 2007). In addition, the centers are tasked with supporting efforts on domestic waste minimization.

Because countries during the treaty negotiations were unable to agree on a specific funding method for the regional centers, article 14 of the Basel Convention merely declares that the parties during COPs should decide on an appropriate and voluntary funding mechanism. Relying merely on voluntary funding for these centers, however, has made it difficult to raise the necessary resources to meet regional demands on services from all regional centers (and those demands will increase if additional centers are established). Funding for the centers continues to be controversial (Earth Negotiations Bulletin 2004a, 2006a, 2008a). Despite verbal commitments from a few Northern donor countries and efforts to secure resources from within each region, many developing countries argue that the regional centers are chronically underfunded and understaffed because they are unable to raise funds for effective operation.

The operation and funding of the regional centers are also areas where governance activities under the Basel Convention are increasingly intersecting with efforts under other treaties and management programs. The regional centers were originally intended to aid in the implementation of only the Basel Convention, but many management and capacity-building efforts under several chemicals treaties have come to overlap with those carried out by the centers. In response, the mandate and scope of the centers have been expanded to include, for example, education and training activities under the Rotterdam Convention and the Stockholm Convention. The regional centers located within the UNECE region may also play similar roles in the implementation of the CLRTAP POPs Protocol. Consequently, the centers have become critical to multilevel governance across the chemicals regime.

Discussions about funding methods and needs for the regional centers are influenced as well by similar debates under other treaties. Many developing countries are pushing for more mandatory funding related to specific capacity-building efforts on chemicals and wastes, including those organized through the centers. These demands are typically resisted by donor countries, which favor maintaining current practices of voluntary funding and providing capacity-building funds through IGOs like the GEF. Governments of most industrialized countries are anxious not to set a major policy precedent regarding mandatory funding in one forum because they fear that will increase pressure from developing countries and advocacy groups to agree to similar kinds of funding methods in other policy forums as well. These linkages shape parties' interests and policy efforts in multiple policy forums across the chemicals regime.

Conclusion

The Basel Convention is the primary global treaty for hazardous waste management and minimization (together with related international agreements on hazardous chemicals and regional agreements on hazardous wastes). Despite the fact that the parties have initiated several important legal and organizational developments under the Basel Convention since the early 1990s, global hazardous waste management continues to suffer from several important shortcomings. Issues for improved management include tackling the growing generation of hazardous wastes, as well as continuing to minimize human health and environmental risks from the transport, reuse, recycling, and disposal of hazardous wastes, including risks resulting from the exposure to hazardous chemicals during these activities.

A global ban on the transboundary transport of hazardous wastes is not politically feasible at present, as many countries engage in the profitable waste trade for both disposal and recycling. A complete trade ban is also not economically and environmentally desirable. Safe disposal of hazardous wastes can be costly, and trade across state boundaries can allow countries to deal with waste disposal in the most economically and environmentally sensible way. Yet there is a continuing need to monitor the transport and disposal of hazardous wastes and stimulate improvements in management techniques to reduce associated risks. To this end, the Basel Convention can be an important instrument for monitoring waste transports and disposal, setting environmentally sound management standards, raising awareness and conducting training, and disseminating the latest waste disposal technology.

While some hazardous waste trade among industrialized countries gives rise to concerns, the trade from industrialized countries to developing countries is the most problematic. In many such cases, wastes are shipped long distances, and receiving countries often lack resources for their safe handling and disposal. A large number of waste items originating in industrialized countries, including e-wastes, are also subject to reuse and recycling in developing countries. This creates a multitude of management challenges as items from laptops to ships are broken up in the search for economically valuable components, many of them extremely hazardous. In addition, illegal trade and dumping pose continuing problems all over the world. In all these areas, legal developments of the Basel Convention and associated management activities are linkage with multilevel governance efforts across other parts of the chemicals regime.

5

The Rotterdam Convention and the Trade in Hazardous Chemicals

This chapter analyzes policy and management issues associated with the international trade in chemicals. In particular, developing countries find it difficult to ensure safe handling and disposal of hazardous chemicals, many of which are imported. For example, a recent inspection in parts of Tanzania found over 17 metric tons of obsolete stocks of the highly toxic pesticides aldrin, dieldrin, and toxaphene (Tanzania 2005). Similarly, a survey in Nepal found over 33 metric tones of obsolete stocks of hazardous pesticides (Nepal 2007). Hazardous chemicals are frequently stored close to human dwellings. Spillages and leakages are common and may contaminate drinking water and soils. Most people lack basic awareness of how to handle chemicals safely. In addition, many countries do not have the capacity to effectively enforce domestic regulations. For example, there are only fifteen pesticide inspectors in all of Tanzania (Tanzania 2005).

Examining trade issues, this chapter focuses on the development of a voluntary PIC procedure followed by the creation of the Rotterdam Convention, the main (but not the only) multilateral agreement regulating shipments of chemicals. The Rotterdam Convention gives parties the legal right to refuse the import of chemicals covered by the treaty, mandating that an exporting country notify the importing country and receive its explicit approval before a shipment can occur. The chapter identifies several issues associated with the formation of actor coalitions and policy diffusion that shape the effectiveness of the Rotterdam Convention and the chemicals regime, including: the establishment of a PIC principle for managing trade in chemicals, the creation of a chemical review committee for evaluating additional chemicals for possible controls, the generation and structuring of financial and technical assistance, and efforts to develop mechanisms for monitoring and compliance.

The chapter begins with a discussion of key regulatory and environmental and human health issues related to the trade in hazardous chemicals and how trade issues emerged on the international political agenda. This is followed by an examination of the development of a voluntary PIC mechanism, including how this was shaped by governance and actor linkages with the similar policy effort on hazardous wastes. Next, the chapter analyzes the main reasons behind the decision by the international community to go from a voluntary mechanism to a convention. To this end, the chapter discusses key policy and linkage issues during the negotiations and implementation of the Rotterdam Convention. The chapter ends with a few concluding remarks on the future of the Rotterdam Convention, including how linkages between it and other forums influence regime effectiveness and multilevel governance.

The Trade in Hazardous Chemicals

The international trade in chemicals has increased dramatically among both industrialized and developing countries since the beginning of the chemicals revolution in the 1940s (OECD 2001). Many industrialized countries began to enact national chemicals legislation in the 1960s in response to growing concerns about food safety, human health risks, and environmental quality, as discussed in chapter 3. This introduction of domestic legislation in industrialized countries, however, resulted in a patchwork of overlapping and conflicting regulations. As a result, chemicals firms operating in multiple countries had difficulty meeting these different sets of domestic standards. The diverse regulations and requirements also created obstacles to international commerce. In response, the OECD began to push for harmonization of domestic regulations among its member states in the 1970s (Alston 1978, Lönngren 1992).

Few of the legislative measures in industrialized countries explicitly regulated exports of domestically regulated chemicals, which left workers and farmers in developing countries largely unprotected, as their national governments lagged in creating domestic regulations and management programs. Despite the fact that most chemicals are traded among industrialized countries, it was the North-South trade that generated the most pressure for developing more stringent international regulations in the 1970s and 1980s. This was similar to what was driving efforts to control the trade in hazardous wastes, as discussed in the previous chapter, with notable linkages between the two issues. As in the case of hazardous wastes, the attempt to regulate the trade in hazardous chemicals was pioneered by developing countries and environmental NGOs operating largely within UNEP, FAO, and the WHO.

Increased pesticide use in developing countries starting in the 1940s, particularly those located in tropical regions, saved millions of people from vector-borne diseases and helped farmers to increase their yields. Yet importing countries rarely had access to environmental and human health–related information, including for many pesticides that were heavily regulated in industrialized countries (Schulberg 1979). A range of local management challenges compounded these problems. For example, many developing countries lacked institutional and human resources for effective education, regulation, and enforcement. As a result, leading IGOs and domestic authorities expressed concerns about the ways in which hazardous pesticides were handled and applied by end users who were unable to read warning labels (when they were present) because of illiteracy and who also often lacked access to protective gear.

Emerging information supported these environmental and human health concerns. The WHO estimated in the 1970s that approximately 500,000 people were poisoned by pesticides annually (Paarlberg 1993, Victor 1998). Data also demonstrated that levels of poisonings relative to pesticide use were much higher in developing countries than in industrialized countries because of domestic management deficiencies. The trade in pesticides to developing countries furthermore increased faster than trade among industrialized countries (although the latter remained higher in absolute numbers), suggesting that people in developing countries would continue to be at risk unless management capabilities in their countries were improved. Consumers in industrialized countries also remained exposed to domestically banned pesticides through the "circle of poison," as discussed in chapter 3.

Nevertheless, because of rising food prices and alarm about food shortages, most governments and firms focused on increasing food production rather than pesticide regulation. In fact, many industrialized and developing countries heavily subsidized pesticides to boost food production in response to population growth. As a middle ground between maintaining the status quo of no international trade controls (increasingly seen to involve human health and environmental costs that were too high) and a blanket export ban (neither socially nor economically advantageous), advocates of trade regulations, including many developing countries and NGOs such as Oxfam and the Pesticides Action Network, coalesced around the idea of a PIC system (Paarlberg 1993, Victor 1998). In doing so, they drew inspiration and lessons from concurrent policy developments of hazardous waste trade.

In 1977, the UNEP Governing Council, based on a developing country—led initiative, adopted a resolution stating that hazardous chemicals should not be exported without the knowledge and consent of the relevant authority of the importing country. Following up on this resolution, the UN General Assembly in 1982 called for the establishment of principles for consent by governments of the importing countries prior to the sale of chemicals banned by the exporting country. Since the 1980s, the development of such a PIC scheme has been the main focus of the international community for managing the trade in hazardous chemicals (Pallemaerts 1988, 2003). These legal and political efforts have been coupled with the promotion of integrated pest management, which seeks to minimize the use of pesticides by developing natural alternatives, and with the designing and communicating of more effective measures for pesticide application when chemicals are used.

Creating a Voluntary PIC Mechanism

The history of the Rotterdam Convention dates back to the 1980s with the development of a voluntary PIC scheme (see table 5.1). This early PIC mechanism was based on activities on policy harmonization and regulation within the OECD, UNEP, and the FAO and was linked to the parallel work on regulating the hazardous waste trade (see chapter 4). This created significant governance and actors linkages between the formally independent but politically and practically linked policy developments on trade in hazardous wastes and chemicals. The establishment of the first global PIC scheme on the trade in commercial chemicals was adopted by UNEP and FAO in 1989, but only after significant political differences among states, IGOs, and NGOs were at least partially overcome. While some substances covered by this PIC scheme were banned or severely restricted in most countries, others remained in widespread use.

Table 5.1
Chronology of important events in the creation and implementation of the Rotterdam Convention

Time	Event
November 1985	1985 FAO Council adopts Code of Conduct without a PIC procedure
June 1987	UNEP Governing Council adopts London Guidelines without a PIC procedure
May 1989	UNEP Governing Council adopts Amended London Guidelines with a first voluntary global PIC procedure
November 1989	1989 FAO Council amends code of conduct to include a voluntary PIC procedure
March 1996	First negotiation session of the Rotterdam Convention
September 1998	Rotterdam Convention is adopted, covering twenty-seven chemicals
February 2004	Rotterdam Convention enters into force
September 2004	COP1 held
February 2006	Mandatory PIC list becomes operational with thirty-nine chemicals
October 2008	COP4 adds one substance to the PIC list, making it forty regulated chemicals
July 2009	127 states and the EU have ratified the Rotterdam Convention

In an effort to harmonize national standards, the OECD Council in 1984 recommended that member states implement a collective notification system for trade. Approximately 75 percent of all chemicals trade at the time was among OECD members (Victor 1998). Already part of U.S. domestic legislation, this notification scheme required an exporting country to inform an importing country of domestic restrictions, and it allowed the importing country to request more information before accepting the import. Thus, the proposed OECD system stopped short of a formal PIC-based system because it would not require official consent by the importing country. Nevertheless, OECD work on data harmonization and notification fed into the work on a more comprehensive PIC scheme carried out within UNEP and the FAO with respect to issues such as information sharing, labeling, and hazard communication.

A coalition of developing countries, with support from some industrialized countries such as the Netherlands and Belgium and working with IGOs and NGOs, offered intellectual and material leadership as they called for a PIC scheme giving countries the right to reject shipments. The chemicals industry objected to such a system, however, arguing that it would be administratively burdensome and too trade restrictive. Several industrialized countries with large chemical firms, including the United States, the United Kingdom, Germany, and Japan, expressed similar concerns. The FAO and UNEP also approached the issue from somewhat different perspectives (Paarlberg 1993). The FAO, set up in 1945, had close ties with agricultural agencies and the agrochemical industry, and focused on food production rather than regulating chemicals. During the early discussions establishing a PIC scheme, the FAO often sided with the opponents of stricter trade controls.

In contrast, UNEP, created in 1972 to address environmental issues, was more supportive of expanding controls and cooperated on this issue with developing countries and environmental NGOs. UNEP established the IRPTC in 1976 to collect and disseminate data on chemical hazards. Two years later, the UNEP Governing Council requested that the IRPTC include information regarding regulations on use in exporting countries. The UNEP Governing Council in 1984 furthermore established the Provisional Notification Scheme for Banned and Severely Restricted Chemicals. Although developing countries had hoped that this scheme would include export regulations, as stated in the resolutions by the UNEP Governing Council in 1977 and the UN General Assembly in 1982, the UNEP Governing Council charged the IRPTC only with developing a database of domestic decisions to ban or severely restrict chemicals. Today the information gathering originally carried out by the IRPTC is administered by UNEP Chemicals (Victor 1998, Pallemaerts 2003).

In the early 1980s, pressure from a growing coalition of developing countries and environmental NGOs such as Oxfam and Greenpeace that was pushing for more stringent export regulations became so strong that the FAO, the major chemicals-producing countries, and the chemicals industry were forced to accept a policy change (Paarlberg 1993). As a result, FAO and its coalition partners now endorsed the idea of codifying a voluntary mechanism for information sharing (though remaining steadfast in their opposition to a legally binding instrument). As a result, between

1982 and 1985, FAO drafted a voluntary code of conduct that codified major principles for the trade of pesticides. Yet there were significant differences among stakeholders over which specific principles the FAO code should contain.

While the coalition of developing countries and environmental NGOs advocated strongly for the inclusion of a formal PIC principle in the FAO code, this was rejected by, among others, the United States, United Kingdom, Germany, the chemicals industry, and the FAO (many of which also opposed stricter trade controls on hazardous wastes). Because the members of this coalition could not be convinced to change their position, the FAO Code of Conduct on the Distribution and Use of Pesticides adopted in 1985 failed to contain a PIC procedure. Instead, it clarified the responsibilities of both exporting and importing parties and shifted more responsibility to exporters, but it stopped short of demanding consent for export. It also contained recommendations for governments and industry on issues such as information exchange, regulation, testing, labeling, packaging, and advertising. In addition, the code identified integrated pest management as best practice (Kummer 1999).

Parallel to the FAO work, the UNEP Governing Council in 1982 established a working group to draft a set of more comprehensive guidelines for information sharing, building on the IRPTC. As a result of these deliberations, UNEP Governing Council in 1987 adopted the London Guidelines for the Exchange of Information on Chemicals in International Trade. As during the negotiations of the FAO code, a coalition of primarily developing countries and environmental NGOs pushed for the inclusion of a formal PIC procedure in the London Guidelines. This was once again blocked by the opposing coalition made up of major chemical-producing countries and the chemical industry. Thus, the final draft of London Guidelines also did not include a PIC scheme requiring explicit consent by importing countries, instead relying on the principle of voluntary information sharing and notification, similar to the FAO code (Kummer 1999).

However, demands for a PIC scheme grew as the supportive coalition of developing countries, a few European countries, and major environmental NGOs gained momentum. The North-South division came to a head at the UNEP Governing Council meeting in 1987 (Paarlberg 1993). As a compromise, UNEP Governing Council decided that the London

Guidelines would be adopted without a PIC provision, but that it would be added at its next meeting, in 1989. The supporters of the PIC scheme quickly used this decision as leverage to pass a resolution at the FAO conference in 1987, ordering FAO to add a PIC provision to its code. Importantly, these policy changes were in part shaped by industrialized countries' acceptance of a PIC scheme on hazardous waste trade; having accepted it for the trade in hazardous wastes made rejecting the same principle for managing the trade in hazardous chemicals politically difficult.

The PIC procedure operated in three sequential steps (Paarlberg 1993). First, the government of an exporting country, on behalf of a domestic firm, was required to notify the authorities of an importing country of any regulatory action that it had taken to ban or severely restrict a chemical on the PIC list intended for export. Second, the government of the importing country was obligated to respond to this notification within ninety days, stating whether it accepted or rejected import of the chemical. Third, the FAO (responsible for pesticides) or UNEP Chemicals (managing the industrial chemicals) would disseminate the response from the importing country government to the exporting country government, which was responsible for communicating this response to the firm seeking export permission and ensuring that the firm complied with the decision of the importing country's government.

Governments were responsible for notifying FAO and UNEP Chemicals of any domestic action to ban or severely restrict a chemical, which made it eligible for inclusion on the PIC list. FAO or UNEP Chemicals was responsible for preparing a decision guidance document for each chemical that was covered by the PIC scheme, summarizing published risk assessment data. These documents were sent to all governments to be used as bases for decisions regarding possible future import restrictions or bans. National governments were also required to summarize their decisions in a written importing country response, which FAO and UNEP Chemicals collected and then submitted to all national government agencies that were working with the two organizations so that governments could compare import restriction decisions on a particular chemical across countries.

A group of national experts who met during the FAO/UNEP Joint Meetings on Prior Informed Consent oversaw the PIC list in collaboration with staff from FAO and UNEP Chemicals. Industry organizations and environmental NGOs also attended these meetings. Victor (1998) characterizes the implementation of the PIC scheme as "learning by doing," as governments and stakeholders attempted to create an effective governance mechanism. This mechanism contained information on four kinds of domestic actions: ban (a substance could not be produced or used), severe restriction (a substance could be produced and used only under limited specified conditions), rejection (a substance had been rejected for registration), and withdrawal (a substance had been voluntarily withdrawn by the manufacturer). Chemicals could be both added to and removed from the PIC list as national regulations changed. By 1995, seventeen chemicals were on the list, increasing to thirty-eight by 1997 (Victor 1998).

Negotiating the Rotterdam Convention

Analysts have argued that a major advantage of the voluntary PIC scheme was that it made it possible to move gradually toward stricter standards and regulations than would have been possible if the international community at the time had attempted to create a legally binding mechanism (Paarlberg 1993, Victor 1998). However, demands from a coalition of most developing countries, several environmental NGOs, and a growing number of European countries for converting the voluntary scheme into a treaty to create a stronger legal basis for environmental and human health protection increased steadily in the 1990s. Chapter 19 of agenda 21 called on states to create a mandatory PIC instrument, and the FAO Council (in 1994) and the UNEP Governing Council (in 1995) approved treaty negotiations. The Rotterdam Convention was negotiated between 1996 and 1998 under joint UNEP and FAO auspices.[1]

FAO and UNEP played important intellectual and material leadership roles during the negotiations, for example, by organizing meetings and producing a draft treaty text. Because the negotiations built on a mechanism accepted by most countries, many negotiators hoped that they could conclude a convention with relative ease, but treaty negotiations proved difficult (Kummer 1999, Earth Negotiations Bulletin 1998a). Coalitions led by the EU and the United States, respectively, clashed over several principal issues, and groups of industrialized countries and developing countries argued over others. A few industry organizations and environmental NGOs attended the negotiations, but their presence was relatively limited (Earth Negotiations Bulletin 1997b). Three sets of issues were particularly difficult: the objective and scope of the PIC procedure, the relationship with the World Trade Organization (WTO) and dispute settlement, and issues of financial and technical assistance. Several aspects of these issues were also influenced by governance and actor linkages with parallel discussions in other parts of the chemicals regime.

Objective and Scope of the PIC Procedure
The decision to negotiate a convention raised important issues about its potential objective and scope. Some developing countries proposed a ban on the export of nationally prohibited chemicals from OECD countries to other countries akin to the Basel Ban Amendment on hazardous wastes, which was adopted in 1995 but had not yet entered into force (see chapter 4). That is, they attempted to use the Basel Convention to leverage similar types of regulations under the Rotterdam Convention by diffusing policy ideas across forums. Similarly, a trade ban under the Rotterdam Convention would increase pressure to ratify the Basel Ban Amendment. However, the idea of a ban received little support from most countries. Because of this lack of support, the negotiations largely focused on transforming the voluntary procedure into a mandatory one (Kummer 1999).

Developing countries argued for the inclusion of language that recognized the principle of common but differentiated responsibilities of industrialized and developing countries under the Rotterdam Convention, similar to its inclusion in other major environmental treaties negotiated in the 1990s. This proposal, however, was rejected by industrialized countries that did not want to make any legal separation between different kinds of exporting and importing countries (Earth Negotiations Bulletin 1997b, 1998a). As a result, the primary objective of the Rotterdam Convention is "to promote shared responsibility and cooperative efforts among Parties in the international trade of certain hazardous chemicals in order to protect human health and the environment from potential harm and to contribute to their environmentally sound use" (article 1), without making any distinction between industrialized and developing countries.

Countries disagreed over what kind of pesticides should be covered. While the United States and Australia believed that the treaty should be limited to "acutely hazardous" pesticides, the EU and many developing countries argued that it should also cover pesticides with long-term chronic effects (Earth Negotiations Bulletin 1997a, 1997b). As a compromise, the Rotterdam Convention covers "severely hazardous" pesticides that produce "severe health or environmental effects observable within a short period of time after single or multiple exposure, under conditions of use" (article 2). Thus, a pesticide is eligible for inclusion on the PIC list in annex III of the convention if it can be categorized as "severely hazardous" (article 6). In contrast, an industrial chemical must have been banned (all uses prohibited) or severely restricted (virtually all uses prohibited) in a country in order to be considered for the PIC list (article 5).

The requirements and procedures for inclusion of a chemical on the PIC list were subject to much controversy. These discussions included questions about the number of countries that needed to submit notifications, the number of different regions these countries had to be from, and how to define these regions. Regulatory action in one country was enough for a chemical to qualify for inclusion under the voluntary PIC procedure; however, many countries wanted to make this requirement harder as they moved from a voluntary procedure to a legal mechanism (Kummer 1999). Canada and New Zealand suggested that there would need to be at least five notifications from three of the seven FAO regions to ensure that a chemical was a worldwide problem.[2] In contrast, the EU and many developing countries believed that the PIC procedure should begin with a single notification, as under the voluntary scheme (Earth Negotiations Bulletin 1997a, 1998a).

In the final compromise, the Rotterdam Convention stipulates that a party that has taken regulatory action banning or severely restricting any chemical of a category not explicitly excluded under the treaty is required to notify the secretariat about such action. For a banned or severely restricted chemical to be included on the PIC list, at least one party each from at least two PIC regions must have taken regulatory action and must have submitted individual notifications to the secretariat. Countries were, however, unable to agree on how to define a PIC region, referring that decision to the COP. The notifications must contain basic substance identification data (e.g., its common name, trade name, and code numbers), information on its use areas and chemical properties, and information on the character and extent of national regulatory action (these requirements are specified in article 5 and annex I).

In addition, a single party that is a developing country or a country with an economy in transition can initiate the inclusion on the PIC list of a severely hazardous pesticide if that country is experiencing domestic problems with its use (a single industrialized country cannot do this) (article 6). The reason behind granting a developing country or a country with an economy in transition this right is that if a country with limited domestic management capacities experiences a problem with a particular pesticide, it should be enough to begin a notification process (Kummer 1999). After the secretariat has received a notification, it has to verify that the information requirements on national regulatory measures outlined in article 6 and annex IV are met. In addition, the secretariat is tasked with gathering information on the environmental and human health risks of the proposed chemical and documenting its legal status in other countries.

The secretariat then forwards all the substance information to the chemical review committee, which reviews the proposal. If the committee concludes by at least a two-thirds majority that a chemical should be added to the PIC list, it submits its recommendation to the next COP. Industrialized countries demanded a consensus vote in the COP, arguing that otherwise any change to the PIC list would have to be subject to formal ratification by the parties (Earth Negotiations Bulletin 1997b, 1998a). In contrast, many developing countries preferred majority voting, fearing that mainly industrialized countries would prevent the inclusion of economically valuable chemicals (Earth Negotiations Bulletin 1997a). Because industrialized countries refused to compromise, the COP operates on the basis of consensus. A decision by the COP on adding a chemical to the PIC list comes into force for all parties at the same time.

Furthermore, country opinions differed over the extent of the notification scheme (Kummer 1999). Many developing countries believed that a mandatory and detailed shipment-by-shipment notification scheme was essential to the effectiveness of the Rotterdam Convention (Earth Negotiations Bulletin 1997a, 1997b). In contrast, major industrialized countries such as Australia, Canada, Japan, and the United States supported notification requirements for only the first shipment, also arguing that not all export notifications had to be mandatory (Earth Negotiations Bulletin 1997a). Seeking a middle ground, the EU proposed an annual mandatory notification process detailing the quantity of the chemicals that were shipped (Earth Negotiations Bulletin 1997a). In the end, countries agreed that export notification for a chemical should be provided prior to the first export, followed by export notifications before the first export in any calendar year (article 12).

A party can respond in three ways to an import request, similar to how the voluntary mechanism operated (article 10). First, it can declare that it consents to receive the import and any other shipments within the same calendar year; second, it may reject the request; or third, it may consent to import, but only if the exporting party meets specific conditions. The secretariat acts as facilitator, distributing all the responses among the parties. National governments are responsible for communicating all information and decisions from the other party to domestic firms. Exporting parties are also required to notify importing parties about chemicals that are domestically banned or severely restricted but not yet on the PIC list (article 12). The importing party must acknowledge that it has received the notification, but the treaty does not regulate the response given by the importing party. If the exporting party does not receive a reply to its notification within a specified time period, it can move ahead with the trade.

The Rotterdam Convention stipulates that the COP, in collaboration with the World Customs Organization, should develop harmonized customs codes for all chemicals on the PIC list for easy identification (Earth Negotiations Bulletin 1997a, 1997b, 1998a). Information labels and safety data sheets should also, "as far as practicable," be in one or more of the official languages of the importing country (article 13). Finally, the Rotterdam Convention contains requirements for the removal of a chemical from the PIC list (article 9). This becomes relevant when a party submits new information to the secretariat, demonstrating that the substance no longer meets the criteria in annex II (for industrial chemicals) or annex IV (for pesticides). This information is forwarded from the secretariat to the chemical review committee, which reviews the case and makes a recommendation to the COP for final decision making.

Relationship with the WTO and Dispute Settlement

During treaty negotiations, countries faced the issue of relating obligations under the Rotterdam Convention with international trade law and other WTO rules. This was a highly controversial issue that caused significant tensions. One coalition of countries, led by the United States, called for an article in the treaty that stated that any provisions in the Rotterdam Convention did not affect the rights and obligations of countries under the WTO, echoing similar demands in other treaty negotiations on chemicals and other environmental issues. The EU and many developing countries, however, opposed the inclusion of text that explicitly stated that all activities were to be "in accordance with WTO rules," as they have also done elsewhere (Earth Negotiations Bulletin 1997b, 3). They believed that such a provision would in effect give superiority to the WTO over the implementation of the Rotterdam Convention, a situation they wanted to avoid (Kummer 1999).

In the end, the inclusion of a specific article was rejected. In a compromise between the two opposing coalitions on how to approach trade and environment issues in environmental treaties, the subject was instead addressed in the preamble. The preamble recognizes "that trade and environment policies should be mutually supportive with a view to achieving sustainable development." At the same time, in an implicit reference to the WTO and other trade-related treaties, it states that "nothing in this Convention shall be interpreted as implying in any way a change in the rights and obligations of a Party under any existing international agreement applying to chemicals in international trade or to environmental protection." The preamble also notes that "the above recital is not intended to create a hierarchy between this Convention and other international agreements."

To avoid conflict with WTO rules on nondiscrimination, the Rotterdam Convention stipulates that decisions by an importing party must be trade neutral. If a party refuses the import of a chemical on the PIC list, it must also prohibit all national production for domestic use of that chemical, as well as cease to accept imports from all other countries—both parties and nonparties (article 10). However, the PIC procedure applies only to Rotterdam Convention parties, and the PIC scheme is not mandatory in trade with nonparties, in contrast to the Basel Convention, which covers imports and exports to both parties and nonparties.[3] Therefore, a party can at least theoretically avoid PIC requirements by using a nonparty country as an importer or exporter intermediary, which would not formally violate any legal commitments under the Rotterdam Convention (Emory 2001, McDorman 2004).

In addition, Canada proposed the creation of a relatively strong mandatory dispute settlement mechanism also involving the use of a nonbinding conciliation commission or the International Court of Justice. Such a mechanism would have gone beyond the compliance schemes in most other environmental treaties at the time (Earth Negotiations Bulletin 1998a). A main reason behind Canada's proposal was to avoid referrals of trade disputes to the WTO by having an effective settlement mechanism under the Rotterdam Convention. This proposal gained support from several smaller countries. However, because this was largely uncharted territory in multilateral environmental negotiations and no easy consensus around a mandatory settlement mechanism emerged, most industrialized and developing countries asked for more time to consider this issue, which also appeared in the context of the Stockholm Convention (see chapter 7) (Kummer 1999).

In the end, the Rotterdam Convention, like other treaties under the chemicals regime, did not initially include a mandatory dispute settlement mechanism because of resistance from major countries. Instead, article 20 stipulates only that parties should settle any disputes between them "through negotiation or other peaceful means of their own choice." Article 20 gives parties two options for settling disputes. First, a party that is not a regional economic organization (any but the EU) may declare in writing on ratification that it accepts compulsory arbitration by a body established by the COP (an arbitral tribunal) or the International Court of Justice (paragraph 2). Second, if both parties to a dispute have not accepted compulsory arbitration, the dispute shall be submitted to a conciliation commission developed by the COP at the request of one of the parties, which would render a recommendation (paragraph 6).

Issues of Financial and Technical Assistance
The Rotterdam Convention requires parties to strengthen national capacities to support domestic implementation (articles 4 and 15). To this end, parties are expected to initiate a host of measures, including designating a national authority to perform all administrative duties under the treaty, adopting national legislation in support of the PIC scheme, creating national registers and databases on chemicals information, and establishing procedures for communicating with firms and the public. The Rotterdam Convention furthermore calls for international exchange of information

and requires that parties provide each other with scientific, technical, economic, and legal information on the management of chemicals and domestic regulatory actions (article 14).

Capacity-building issues related to the fulfillment of all these tasks were hotly debated and linked with similar debates under the Basel Convention and the then forthcoming negotiations of the Stockholm Convention (see chapters 4 and 7). Industrialized countries argued that importing countries were primarily responsible for strengthening their own national legal and political structures (Earth Negotiations Bulletin 1997a). Developing countries and countries with economies in transition, however, emphasized the need for financial and technical assistance in support of their implementation of the Rotterdam Convention, arguing that they otherwise would not be able to meet their commitments. Among the challenges that many developing countries faced were meeting administrative demands on importing countries under the PIC scheme, improving measures to combat illegal trade, and enhancing management capabilities for obsolete chemicals (Earth Negotiations Bulletin 1997b).

Funding discussions were also influenced by similar debates in other forums. Developing countries called for the creation of a separate financial mechanism modeled after the Multilateral Fund of the Montreal Protocol. Countries such as Malaysia and Algeria also suggested that industry provide funding for capacity building (Earth Negotiations Bulletin 1997a). These proposals, however, were rejected by industrialized countries, which were willing to accept only voluntary commitments. In part, their refusal to accept any mandatory commitments was shaped by their desire not to set any precedents for funding that could be used as leverage under other treaties, both within and outside the chemicals regime. As a result, the Rotterdam Convention merely states that parties "with more advanced programmes for regulating chemicals should provide technical assistance, including training, to other Parties in developing their infrastructure and capacity to manage chemicals throughout their life-cycle" (article 16).

Implementing the Rotterdam Convention

The Rotterdam Convention establishes a legal framework for addressing the import and export of chemicals. The treaty entered into force in 2004, and 127 countries and the EU were parties as of late 2009. The United States and Russia are two notable nonparties. During the six years between the adoption and the entry into force of the convention, the PIC procedure was run on a voluntary basis based on the Resolution on Interim Arrangements passed by countries alongside the adoption of the treaty (Earth Negotiations Bulletin 1998b, McDorman 2004). In addition, six more meetings of the international negotiating committee were held between 1999 and 2003 to oversee the operation of the voluntary PIC procedure and prepare for treaty implementation, including the launching of the mandatory PIC scheme.[4]

Several contentious issues were dropped from the treaty negotiations in order to conclude a convention in a timely manner (Earth Negotiations Bulletin 1998a, Kummer 1999). Nevertheless, countries left these issues off the agenda only temporarily. In particular, the COPs have focused much attention on three sets of important implementation issues: organizational and legal developments, implementing and expanding the mandatory PIC procedure, and financial assistance and technology transfer.[5] Many of these legal, political, and practical issues are also linked with debates and policy making under the Basel and Stockholm conventions in particular, as the implementation of these three treaties (alongside the CLRTAP POPs Protocol) intersects with and shapes the behavior of regime participants and policy outcomes across forums.

Organizational and Legal Developments

The Rotterdam Convention is administered by a secretariat organizationally divided between UNEP Chemicals in Geneva and FAO in Rome. Parties early on specified principles for arbitration and conciliation using the dispute settlement mechanism, as outlined in article 20 and annex VI of the convention. Decisions of the arbitral tribunal, which are taken by a majority vote, are binding on the two parties before the tribunal, as well as any intervening third party. In contrast, parties are required to follow only the recommendations of the conciliation commission, which are also taken by a majority vote, "in good faith" (that is, parties are not legally required to follow them but can be expected to come under political pressure to do so).

The establishment of a compliance mechanism to work alongside the dispute settlement mechanism is another major issue that was left out of the Rotterdam Convention, and it is one where the parties so far have failed to reach agreement (Earth Negotiations Bulletin 2008b). As under the Basel Convention, compliance issues are highly controversial, as the opinions of country coalitions differ noticeably. Parties are also aware that any major decision that they take under the Rotterdam Convention is likely to have ramifications for how compliance issues are addressed in other forums, including under the Stockholm Convention. This stalls progress across policy forums. Key issues include how the compliance mechanism would operate; the regional composition of the compliance committee; and the use of measures against noncompliant parties, including sanctions.

Major countries, including Australia, China, India, Japan, and the United States, favor a relatively weak compliance mechanism with few punitive powers. In contrast, the EU, Canada, New Zealand, and several smaller developing countries support a mechanism that has the ability to punish noncompliance. Several countries have stated that they would accept only a self-trigger arrangement. Their opposition to party-to-party and secretariat submissions is shaped by their belief that the Basel Convention compliance mechanism, which accepts those kinds of submissions, is too aggressive (see chapter 4). In particular, many developing countries also resist allowing NGOs to submit complaints, as Germany proposed (Earth Negotiations Bulletin 2002b, 2005a). As parties continue to work on these issues, their resolution will also be shaped by continuing discussions on compliance issues under the Basel and Stockholm conventions.

Implementing and Expanding the Mandatory PIC Procedure

The Resolution on Interim Arrangements created the interim chemical review committee to oversee the PIC procedure until the Rotterdam Convention entered into force.[6] To this end, countries had to address several critical issues regarding the membership and operation of the interim committee (Earth Negotiations Bulletin 1999b). At COP1 in 2004, the parties agreed to extend the interim procedure until February 14, 2006, done for largely practical purposes (Earth Negotiations Bulletin 2000a). Over 160 countries participated in the voluntary PIC scheme. The Rotterdam Convention needed only fifty ratifications to enter into force, which meant that the mandatory mechanism would have dramatically fewer members if it immediately replaced the voluntary one. It was therefore agreed that the interim mechanism would operate beyond the time that the Rotterdam Convention was expected to enter into force.

The interim chemical review committee consisted of twenty-nine experts from all seven FAO regions. Most industrialized countries believed that the chemical review committee should operate the same way. At COP1, however, the debates of what constituted "equitable geographic distribution" and a fair balance in membership between developed and developing countries were intense, mirroring parallel debates under the Stockholm Convention (see chapter 7). Proposals from mainly developing countries for changing the composition of the chemicals review committee included making the number of experts proportionate to the number of parties from each region and enhancing the representation of developing countries and countries with economies in transition. Proponents of these proposals argued that they were needed to enhance the salience and legitimacy of the review committee (Earth Negotiations Bulletin 2004b, Kohler 2006).

Most proposals by developing country to change the way the chemicals review committee would be staffed and operate were opposed by industrialized countries, which wanted to cap the size of the committee for budgetary reasons. Industrial countries also argued that a smaller group would operate more effectively while still be able to produce usable knowledge and inform policy making (Kohler 2006). The final compromise, which left all parties "equally unhappy," resulted in a chemical review committee with thirty-one expert members: eight from Africa, eight from Asia and the Pacific, seven from Western Europe and others, three from Eastern Europe, and five from Latin America and the Caribbean (Earth Negotiations Bulletin 2004b, Kohler 2006). Committee membership, which is drawn from the five traditional UN regions rather than the seven regions most commonly used by the FAO, rotates every two years. Experts serve four-year terms for a maximum of two consecutive terms.

Another much-debated aspect of the operation of the chemicals review committee concerned language issues. While meetings of the interim review committee were conducted only in English, several developing countries at COP1 called for the simultaneous interpretation of the meetings of the chemical review committee into all six official UN languages. They argued this was necessary to ensure full participation of experts from non-English-speaking countries. Industrialized countries, however, believed that the cost of interpretation services would be too high. Ending in a halfhearted compromise to meet the demands of the French- and Spanish-speaking delegations in particular, COP1 decided that the meetings of the chemical review committee would be carried out in a single language but without specifying which one (Kohler 2006). This made English the effective, if not the formal, language of the review committee.

Both industry organizations and environmental NGOs regularly observed meetings of the interim chemical review committee and also continue to be involved with the chemical review committee (Kohler 2006). Often different sets of industry groups and firms dominate among the stakeholder observers. Which associations and firms that attend a meeting depends partly on which specific chemicals are under review. Among the environmental NGOs, especially the Pesticide Action Network is closely following policy developments and is also actively involved in identifying problems with hazardous pesticides with partner organizations in developing countries. The Pesticide Action Network also cooperates with labor organizations such as the International Union of Food, Agricultural and Allied Workers' Associations on developing programs for collecting and analyzing data on PIC chemicals to be presented to governments and IGOs for the purpose of shaping debates and policy decisions (Earth Negotiations Bulletin 2000a).

The chemical review committee first met in 2005, one year before the mandatory mechanism entered into force.[7] Twenty-seven substances were originally controlled by the Rotterdam Convention, carried over from the voluntary scheme. COP1 added fourteen chemicals following recommendations by the interim chemical review committee (Earth Negotiations Bulletin 2005a). In addition, COP4 included tributyltin compounds based on the review by the chemical review committee (Earth Negotiations Bulletin 2008b). The PIC procedure covered forty chemicals as of 2009 (see table 5.2). Although twenty-seven substances plus fifteen substances equals forty-two substances, only forty substances are on the PIC list. This is because when COP1 added all formulations of monocrotophos and parathion, some of these were already listed separately. As some individual listings were merged, there were thirty-nine separate chemicals on the PIC list. COP4 added one more substance, and no chemicals have been removed from the PIC list.

Table 5.2

The 40 chemicals covered by the Rotterdam Convention by 2009

2,4,5-T	Ethylene oxide	Phosphamidon
Aldrin	Fluoroacetamide	Methyl-parathion
Binapacryl	HCH (mixed isomers)	Crocidolite
Captafol	Heptachlor	Actinolite
Chlordane	Hexachlorobenzene	Anthophyllite
Chlordimeform	Lindane	Amosite
Chlorobenzilate	Mercury compounds	Tremolite
DDT	Monocrotophos	Polybrominated biphenyls (PBB)
Dieldrin	Parathion	PCB
Dinitro-ortho-cresol (DNOC)	Pentachlorphenol	Polychlorinated terphenyls (PCT)
Dinoseb	Toxaphene	Tetraethyl lead
1,2-dibromoethane (EDB)	Dustable powder formulations containing a combination benomyl, carbofuran and thiram	Tetramethyl lead
Ethylene dichloride	Methamidophos	Tris (2,3-dibromopropyl) phosphate
Tributyltin compounds		

During the operation of the voluntary scheme, countries generally followed the recommendations of the FAO/UNEP and the interim chemical review committee for adding chemicals to the PIC list. As another 160 chemicals are lined up for review by the chemical review committee (Earth Negotiations Bulletin 2006b), there is a growing unease among parties that the PIC procedure is becoming increasingly contentious and politicized, a view supported by recent events. The chemical review committee at its first meeting rejected thirteen of the fourteen chemicals notified by parties, arguing that the notifications lacked sufficient risk evaluation data (Earth Negotiations Bulletin 2005a). Furthermore, two chemicals have become highly controversial as their inclusion on the PIC list has been blocked by minority coalitions despite recommendations for listing by the chemical review committee: chrysotile asbestos and endosulfan.

A small group of countries led by Russia and Canada throughout the first four COPs led the opposition against including chrysotile asbestos (Earth Negotiations Bulletin 2004b, 2005a, 2006b, 2008b). At COP4, the listing of endosulfan was prevented by another small group of countries led by India, the world's largest producer, user, and exporter of this substance (Earth Negotiations Bulletin 2008b, Tupper 2009). The discussions on endosulfan are also shaped by equally contentious debates on endosulfan under the Stockholm Convention (see chapter 7) and the CL-RTAP POPs Protocol (see chapter 6). That is, the treatment of endosulfan under each of these treaties affects policy decisions under the other two. Because of these kinds of conflicts, many parties fear that at least some countries will become more restrictive in their preferences about listing chemicals on the mandatory PIC list. Statements by several developing countries also indicate that they view the PIC list as a de facto blacklist of chemicals that should not be subject to any trade (Earth Negotiations Bulletin 2004b).

Identification issues are furthermore important to the operation and effectiveness of the PIC procedure. The secretariat and the parties work with the World Customs Organization on developing a collective system of customs codes for the chemicals on the PIC list (Earth Negotiations Bulletin 2000a, 2004b). The Rotterdam Convention also operates alongside the Globally Harmonized System for the Classification and Labeling of Chemicals, a voluntary system adopted in 2003 that establishes criteria for classifying chemicals and develops compatible labeling and safety data sheets. Countries without domestic systems for hazard classification and labeling are urged to adopt the global criteria, and countries that already have domestic systems are expected to align them with the global criteria. UNITAR, ILO, and OECD are the main organizations working with developing country governments on criteria and labeling issues.

For the PIC scheme to work as intended, possible importing parties must respond to the trade requests submitted through the secretariat. Yet many countries lack the human and technical resources to review all requests and adequately control imports. Statistics developed by the secretariat showed the overall response rate was only 50 percent for requests generated through the voluntary mechanism. The response rate furthermore fell to below 30 percent for chemicals added to the PIC list after 1998 (McDorman 2004). In a case where there is no response, a complex set of legal rules outlines the continued responsibility of the potential exporter. However, a high degree of nonresponses clearly weakens the effectiveness of the Rotterdam Convention. Of course, this problem will become bigger the more substances that are added to the PIC list. This situation, together with other concerns, raises critical issues about financial and technical assistance.

Financial Assistance and Technology Transfer

The Rotterdam Convention, unlike the Basel and the Stockholm conventions, does not include any provisions for a financial mechanism. Many developing countries and countries with economies in transition were bitterly disappointed that financing issues were largely left out of the treaty (Earth Negotiations Bulletin 1998a, 1998b). The Resolution on Interim Arrangements called for voluntary technical assistance to developing countries but offered few specifics. It was decided that developing countries in need of assistance would send their requests to the secretariat, which would communicate these requests to possible donor countries (Earth Negotiations Bulletin 2000a). Developing countries continue to push for the creation of a permanent treaty-specific financial mechanism (Earth Negotiations Bulletin 2004b, 2006b). Yet industrialized countries remain opposed to a mandatory and treaty-specific financial mechanism. Instead, they prefer linking the activities of the Rotterdam Convention to existing financial mechanisms.

Mirroring discussions under the Stockholm Convention, industrialized countries in particular like to use the GEF (see chapter 7) (Earth Negotiations Bulletin 2005a, 2006b). Developing countries, however, fear that the GEF is not able to distribute enough funds for effective implementation of the Rotterdam Convention, given the many areas in which it is active. In a compromise that left most developing countries unhappy, parties agreed to set up a voluntary special trust fund under the Rotterdam Convention and further explore how the Rotterdam Convention may pursue additional funding from external sources including those associated with the Stockholm Convention, the GEF, and SAICM (Earth Negotiations Bulletin 2006b). This also includes the use of regional centers under the Basel and the Stockholm conventions, linking capacity-building efforts across multiple treaties. These issues are critical for the implementation of effective multilevel governance efforts within many parts of the chemicals regime.

Conclusion

One observer of the voluntary scheme that existed before the Rotterdam Convention noted that "if it operates properly, PIC could be one of the largest organized transfers of useful regulatory information to developing countries, which in turn could contribute to the ultimate goal of the PIC system—improving management of hazardous chemicals and pesticides" (Victor 1998, 247). It was of great advantage to the creation and early implementation of the Rotterdam Convention that most countries were already familiar with key aspects of the PIC procedure because of their work under the voluntary mechanism for over a decade. However, the effective implementation of the Rotterdam Convention is dependent on a successful solution to several challenges, both old and new.

One critical issue is the continuing expansion of the PIC list, which is marred by political controversy. The effectiveness of the Rotterdam Convention also depends on the ability of developing countries to manage imports. This includes the capacity of customs officials to identify every shipment of a chemical on the PIC list and intercept all those that are not PIC compliant—a daunting task in any country. Given the lack of domestic resources in many developing countries and the economic importance and complexities of the trade in hazardous chemicals, funding and other kinds of assistance for education and domestic capacity building is a critical issue (Emory 2001). Debates on financial and technical support under the Rotterdam Conventions are linked with similar discussion under the Basel and the Stockholm conventions, including the role of GEF and regional centers.

Furthermore, the development of an efficient compliance mechanism is important for overseeing implementation and enhancing effectiveness. This issue is closely related to the discussions on financial support and capacity building. Developing countries argue that it would be inappropriate to decide on a strong compliance mechanism without first creating a reliable financing mechanism (Earth Negotiations Bulletin 2005a); their ability to improve compliance, including with the provisions of completing notifications and adequately responding to export requests, is partially dependent on the amount of external financial and technical aid made available to them for improving domestic management capabilities. Like the finance and capacity-building issue, discussions and decisions on compliance issues are politically and practically interlinked with similar debates under other treaties.

In addition, the Rotterdam Convention may help to improve situations related to the international trade in hazardous chemicals, but it will not solve all problems. Even as there are many hazardous chemicals that are extensively traded that are not on the PIC list, it is possible that countries are becoming more hesitant to include new chemicals on the PIC list because of the stricter legal responsibilities under the Rotterdam Convention compared to the earlier PIC scheme. It is also likely that chemicals that will be proposed in the future are more economically valuable than many of those that were listed in the past, which may cause more intense conflicts among countries. The PIC procedure is moreover not compulsory when a party trades with a nonparty: in many cases, hazardous chemicals can still be legally traded with few requirements. Finally, it is to be expected that there will be continuing illegal trade in regulated chemicals.

6

The CLRTAP POPs Protocol and Regional Standards

This chapter examines regional efforts in the northern hemisphere to regulate POPs. In the early 2000s, an Arctic Monitoring and Assessment Programme AMAP assessment report (2004) concluded that many Arctic species at the top of complex food webs carry high levels of a multitude of POPs. These POPs reach the Arctic mainly through long-range transport, but the report also noted cases of contamination from local industrial activities. Researchers have documented adverse immunological, behavioral, and reproductive effects in wildlife linked to chemical contamination throughout the Arctic (Reiersen, Wilson, and Kimstach 2003). In addition, AMAP reports have expressed concern for human health and, in particular, health risks to indigenous populations. In fact, some of the highest human concentrations of POPs have been detected in people living in Arctic local communities. Political and scientific concerns about this situation fed into much of the development of regional and global POP controls beginning in the 1990s.

Analyzing the emergence of POPs on the international scientific and political agenda, this chapter focuses on the creation and implementation of the CLRTAP POPs Protocol, the first multilateral treaty specifically on POPs (although earlier treaties in part cover chemicals now classified as POPs). The principal objective of the parties' actions on POPs is to control, reduce, or eliminate discharges, emissions, and losses of POPs. Analyzing the creation and implementation of the CLRTAP POPs Protocol, this chapter highlights issues associated with the formation of actor coalitions and policy diffusion influencing the protocol, as well as the effectiveness of other policy developments under the chemicals regime. Such issues include: the scientific and political framing of the POPs issue, the assessment and development of management options for specific POPs, and the creation of a review committee to evaluate additional chemicals for possible controls.

The chapter begins with an examination of the scientific detection of long-range transport of POPs. Next follows an analysis of how emerging scientific data on long-range transport of POPs and associated human health and environmental concerns fed into domestic and international policy making on hazardous chemicals in the 1980s. This is continued by an analysis of the international environmental assessments and negotiations under the UNECE that resulted in the development of the POPs Protocol from 1989 to its adoption in 1998. The following section examines measures to implement the POPs Protocol since 1998. The chapter ends with a few comments on major issues related to implementation of the POPs Protocol and how these are tied to other policy instruments and the effectiveness of the chemicals regime.

Long-Range Transport of Hazardous Chemicals

CLRTAP was adopted in 1979 under the auspices of UNECE, and it entered into force in 1983. As one of the UN's several regional organizations, UNECE was established in 1947 as a subsidiary body of the United Nations Economic and Social Council to facilitate the economic reconstruction of postwar Europe. As a result of the outcome of World War II, UNECE and CLRTAP came to comprise all of Europe (both east and west) as well as the United States and Canada (Chossudovsky 1989).

As a framework convention, CLRTAP established broad principles, working procedures, and an organizational setting as a basis for cooperation on technical, scientific, and policy issues for the reduction of transboundary air pollution within the UNECE region. The convention's primary work tasks include coordinating information gathering and dissemination, overseeing collaborative research programs, facilitating political negotiations on pollution-reduction protocols, and monitoring and reviewing national implementation of programs and commitments (Wettestad 2002, H. Selin and VanDeveer 2003). CLRTAP began addressing hazardous chemicals in the late 1980s. Consistent with the aim of the framework convention, the CLRTAP POPs Protocol focuses on issues relating to long-range transboundary transport of POPs within the UNECE region.

Actions on POPs under CLRTAP are closely linked to scientific and political concerns about Arctic pollution. The Arctic was long viewed as too remote from industrial societies to be at risk of serious contamination. In contrast, it is now recognized that the sensitive Arctic environment is the first to react to environmental hazards and therefore functions as a window to the future. Inuit communities in Greenland and Canada face among the highest exposure to POPs. Negative health effects on for example early brain development and function may be the result of this exposure. Indigenous peoples' exposure to POPs is mainly a result of dietary intake of contaminated Arctic wildlife, often exceeding national guidelines (AMAP 2003, Kuhnlein et al. 2003, Kuhnlein and Chan 2000). Taking into consideration the physiological and nutritional benefits of traditional food systems, as well as their social and cultural importance, dietary recommendations have been developed to minimize human exposure to POPs.

As northern countries introduced domestic controls on a small set of hazardous chemicals beginning in the 1960s, authorities believed they were gaining control over chemical pollution and contamination problems. This optimism was supported by local scientific studies in the 1970s and 1980s that showed a decline in environmental concentrations of several hazardous chemicals and signs of recovery in affected wildlife. By the late 1980s, however, unanticipated scientific data from the Arctic raised new concerns (see chapter 3). This discovery of a wide range of hazardous chemicals in areas far removed from large-scale industrial and agricultural activities was not the first in the Arctic environment—the existence of such substances had been measured since the late 1960s—but the contamination levels that were discovered in the late 1980s were surprisingly high to most scientists and regulators.

In response to this discovery, Canadian authorities, mainly through Indian and Northern Affairs Canada (INAC), undertook a multidisciplinary review of scientific findings of Arctic pollution (Shearer and Han 2003). Aided by improvements in measurements, data samples, and analytical techniques, three connected factors were revealed: systematic long-range atmospheric transport of emissions to the Arctic, high environmental contamination levels throughout the Arctic region, and actual and potential environmental and human health implications. These findings related to a specific category of hazardous persistent organic substances with similar chemical characteristics. Chemicals displaying these POP characteristics included widely used substances such as DDT, PCBs, toxaphene, chlordane, hexachlorocyclohexane (HCH), endrin, dieldrin, hexachlorobenzene, and mirex.

Seeking International Action

Although the Arctic scientific findings on POPs triggered the launch of a wide range of scientific studies in the 1980s to examine emission sources, transport patterns, and environmental and human health risks, they sparked only limited political reaction in Canada. In fact, INAC was the only major Canadian government agency to seek political action at the time based on the new science. In doing so, INAC, because of its area of operation, expressed particular concern for the indigenous populations as it collaborated closely with the ICC and local indigenous communities on domestic and regional scientific and policy issues (N. Selin and H. Selin 2008). These concerns were related to scientific data demonstrating alarmingly high levels of hazardous chemicals in breast milk and blood samples taken from indigenous peoples (Dewailly and Furgal 2003, UNECE 1994).

The scientific data on frequent long-range transport of POPs led INAC officials to believe that international regulatory measures were needed (H. Selin and Eckley 2003). For that purpose, INAC representatives went forum and scale shopping for an organization supportive of its efforts. Initially INAC contacted global organizations such as the OECD, FAO, and UNEP, but they showed little interest in pursuing the issue beyond existing information-gathering mechanisms. Instead, INAC directed their political efforts toward CLRTAP. Until the late 1980s, the primary focus of CLRTAP had been on acidification and eutrophication abatement. Although CLRTAP was more regional in scope than the organizations that INAC first contacted, Canadian officials believed that CLRTAP's experience with northern long-range transboundary air pollution abatement made it a relevant forum for initiating multilateral action.

An INAC official's presentation of a Canadian report on Arctic problems of toxic, persistent, and bioaccumulating chemicals in 1989 persuaded the CLRTAP Working Group on Effects, the body managing assessment activities, to include the issue of hazardous organic substances in its work plan. To gain a better understanding of the scope of the problem, the Working Group on Effects requested that a group of government-designated experts, under Canadian supervision, prepare a report on the effects of hazardous airborne organic compounds, including their sources, transportation, biological uptake, and ecosystem and health effects. The final report was to be considered by the Working Group on Effects in 1991 (UNECE 1989). Arctic cooperation on hazardous chemicals at the same time was also expanded under AMAP and the Arctic Council (N. Selin and H. Selin 2008).

Canada was not the only country taking action on these substances. The Swedish Environmental Protection Agency (SEPA) in the late 1980s also expressed concern over the long-range transport of hazardous chemicals. Environmental concentrations of some chemicals in the Baltic Sea region were leveling off rather than decreasing, while concentrations of other chemicals were rising (Larsson and Okla 1989). In response, SEPA undertook an assessment of the Baltic region similar to that done by INAC for the Canadian Arctic, and reported similar results: evidence of inflow of emissions through long-range atmospheric transport, slowing rates of decline of environmental levels of DDT and PCBs, and rising concentrations of chlordane and other hazardous organic substances. SEPA expressed concern regarding the environmental and human health aspects of these trends (Swedish Environmental Protection Agency 1990).

Based on their shared interests, INAC and SEPA played crucial intellectual and material leadership roles by raising the issue of long-range transport of hazardous persistent organic substances. At the meeting of the CLRTAP Working Group on Effects in 1990, Canada presented a draft of the paper requested a year earlier. Furthermore, Sweden proposed forming a more formal task force to assess the extent and severity of the POPs problem. The proposal was approved by the CLRTAP Executive Body in 1990, the convention's main decision-making body (UNECE 1990). This marked the beginning of a series of assessments and political meetings leading to the adoption of the POPs Protocol (see table 6.1). The CLRTAP POPs assessment work through governance and actor linkages also had significant impacts on other policy and management activities, especially the Stockholm Convention (see chapter 7).

Table 6.1
Chronology of important events in the creation and implementation of the CLRTAP POPs Protocol.

Time	Event
August 1989	First CLRTAP activities on POPs
December 1990	Executive body establishes the task force
June 1994	Task force presents its final report
December 1994	Executive body forms the preparatory working group
March 1995	First meeting of the preparatory working group
October 1996	Final meeting of the preparatory working group
January 1997	Protocol negotiations begin
February 1998	Final negotiation session
March 1998	Executive body takes a decision on requirements and procedures for adding of substances (decision 1998/2)
June 1998	CLRTAP POPs Protocol is adopted
October 2003	CLRTAP POPs Protocol enters into force
July 2009	Twenty-eight states and the EU have ratified the CLRTAP POPs Protocol

CLRTAP POPs Assessments

The CLRTAP POPs assessments had a notable impact on the framing of the POPs issue, shaping both regional and global political actions. These CLRTAP assessments were carried out alongside the initiation of more extensive circumpolar chemicals assessment work under AMAP. A small Task Force on POPs operating between 1990 and 1994 carried out the early CLRTAP assessment work on POPs and issued an assessment report. The task force focused on generating a scientific overview of the POPs problem in the Northern Hemisphere, but also began exploring possible abatement options. The work of the task force was followed by additional assessments conducted by an Ad Hoc Preparatory Working Group on POPs in 1995 and 1996. The working group built on the task force findings as it assessed the scope of the POPs problem in the northern environment and developed specific policy options.

The Task Force
The Task Force on POPs met four times.[1] Most of the work was carried out by a small group of well-networked government officials. Canada

and Sweden continued to display intellectual and material leadership as they cochaired the task force as well as organized and funded meetings and studies feeding into the report. Other active countries included Germany, the Netherlands, Norway, United Kingdom, and the United States. No representatives of the chemicals industry or environmental NGOs participated in the assessments, as the task force operated in relative political anonymity. The task force divided its work into two areas. First, it assessed the physical characteristics of the POPs problem by surveying the scientific literature, focusing on data on emission sources, long-range transport, and distribution among environmental media. Second, it identified and evaluated possible policy responses (H. Selin 2000).

Significantly, the task force constructed POPs as a combined scientific and policy issue. In doing so, the task force established the term *POP* (H. Selin and Eckley 2003). Before the task force, the term *POPs* was not used in scientific or policy circles. Instead, the set of chemicals was described by their basic characteristics. As the task force adopted the more manageable POPs acronym, it quickly made it into scientific papers and external policy discussions in the 1990s. This was facilitated by the fact that many country officials on the task force were active in both scientific and policy communities, describing their work on POPs and commissioning national and international scientific studies on it. As such, the CLRTAP assessments were influential far beyond CLRTAP by generally establishing what is now meant by a POP and identifying POPs as a subset of hazardous chemicals warranting regulatory action.

In framing the POPs issue, the task force agreed on certain physical, chemical, and biological characteristics that constituted a POP (UNECE 1994). This definition, which has carried over into many other policy forums, characterized POPs as a group of hazardous organic chemicals that are persistent in the environment, leading to their bioaccumulation in fatty tissues. Their chemical characteristics also enable them to travel long distances in the atmosphere before deposition. The task force divided suspected POPs into three categories for assessment purposes, which have largely remained intact in subsequent regional and global work, including under the Stockholm Convention: pesticides, industrial chemicals, and by-products. The task force also worked on establishing screening criteria, collecting risk assessment data, and selecting and conducting screenings of substances.

The task force found several suspected POPs to be extensively used, although incomplete records of domestic production and use made exact reviews impossible. The task force report identified a number of important emission sources, including agricultural use, manufacturing and use of goods, spills and dumping, waste incineration, and combustion processes. Studies showed that environmental concentrations were typically highest close to emission sources, but data showed extensive long-range transport of emissions, causing significant environmental accumulation in remote regions that had no local emission sources. Although some hazardous industrial chemicals and pesticides were regulated in a number of countries, their long-range transport continued to be a problem because of incomplete regional control measures on use and disposal. Domestic controls of by-products were less comprehensive (UNECE 1994).

In line with the earlier findings that prompted INAC and SEPA to seek international action and approach CLRTAP, the task force identified the atmosphere as the primary transport medium for POPs. The physical and chemical properties of POPs, in combination with predominant global atmospheric circulation patterns, caused systematic migration of POPs to cooler latitudes. Several negative effects in wildlife were linked with POPs, including immune and metabolic dysfunction, reproductive anomalies, behavioral abnormalities, and carcinogenic effects. Although human data were sparse, available information was consistent with effects reported in both laboratory and wildlife studies. The task force concluded in its final assessment report in 1994 that the weight of evidence "clearly indicates that action to address POPs is warranted now" (UNECE 1994, para. 92).

Furthermore, the task force surveyed existing domestic and international regulatory and management activities on POPs. This survey was based on the premise that effective controls on POPs could be achieved only by targeting their long-range transport from a wide range of emission sources. The task force found that domestic controls were incomplete and also concluded that the other international regimes and organizations it surveyed were unsuitable for achieving comprehensive POPs reductions because their geographical domain, membership, and mandates were too narrow.[2] As a result, the task force found CLRTAP, based on its geographical coverage and its ability to address transboundary air pollution, to be the most appropriate international mechanism for developing controls for POPs. Therefore, the task force recommended the creation of a CLRTAP POPs agreement (UNECE 1994).

To move forward on POPs, the task force suggested a two-track approach because it believed that dealing with all potential POPs at once would be too difficult. Task force members also believed this would strengthen the long-term effectiveness of an agreement. A similar two-track approach was later embedded in the Stockholm Convention, in large part because of its use under CLRTAP. The first track was to achieve a quick protocol on a small list of substances that had already been acted on in many UNECE countries or for which there was ample risk information. Agreeing on measures under track 1 proved an arduous task, however, because opinions diverged on what constituted sufficient risk information and appropriate measures. The second track consisted of establishing a mechanism for the future screening of additional substances for possible controls under the protocol once it entered into force (UNECE 1994).

The final task force report was presented to the CLRTAP Executive Body in 1994. The executive body was generally supportive of the task force's conclusions and agreed that work on POPs should continue under CLRTAP. Nevertheless, states' opinions diverged on how best to proceed (H. Selin 2000). A coalition of a few European states favored starting protocol negotiations based directly on the task force's assessment work; however, a larger coalition of European and North American countries argued that further assessments, particularly with regard to developing more detailed policy alternatives, were needed before protocol negotiations could start. For that purpose, the executive body set up the Ad Hoc Preparatory Working Group on POPs to work under the Working Group on Strategies and Review, the CLRTAP political negotiating committee where all CLRTAP protocols are negotiated.

The Preparatory Working Group

The Ad Hoc Preparatory Working Group on POPs met four times between 1995 and 1996.[3] While more countries attended the this group than the task force, the same relatively small group of European and North American countries continued to dominate the assessment work. Representatives from UNEP Chemicals, by then active in the global POPs work, attended all meetings. Industry organizations, including the

International Chamber of Commerce, the Chemical Manufacturers Association (later renamed the American Chemistry Council), and CEFIC were regularly present. The International Union for Conservation of Nature and Greenpeace also attended a few meetings. The preparatory working group focused on three sets of issues: establishing assessment criteria and screening substances for initial inclusion in the protocol, identifying control options, and developing a mechanism for future evaluation of additional substances (H. Selin 2000).

The preparatory working group established a model—the so-called modified task force methodology—to screen for substances that were subject to long-range atmospheric transport within the CLRTAP region and posed risks to wildlife and human health (focusing on issues of persistence, bioaccumulation, and toxicity) (H. Selin and Hjelm 1999, Rodan et al. 1999). After screening 107 chemicals, the preparatory working group identified fourteen POPs as primary candidates for initial regulation. Yet countries expressed diverging opinions on how the results of the screening should be interpreted and what constituted acceptable risk levels. As a result, some countries insisted that four additional substances—short-chain chlorinated paraffins (SCCP), heptachlor, chlordecone, and lindane—should be included in the protocol negotiations, totaling eighteen chemicals (see table 6.2) (United Kingdom 1996).

The preparatory working group reached an understanding that controls should be mandatory, with regulated substances listed in separate annexes. Considering control options for pesticides and industrial chemicals, countries focused on production, use, import, export, stockpiles, and wastes. Discussions on production and use concerned how bans and exemptions would be applied to individual chemicals. While many European countries believed that effective POPs controls had to include trade restrictions, the United States and Canada argued that it would be inappropriate for a regional agreement to impose import and export controls. Similarly, European countries argued in favor of mandatory controls on the transnational transportation of stockpiles and wastes; however, Canada and the United States viewed these as inappropriate trade restrictions in a regional forum and argued that regulations on trade were best left to the Basel Convention (see chapter 4).

Addressing the release of POPs generated as by-products, the preparatory working group identified major mobile and stationary emission

Table 6.2
Substances identified by the Modified Task Force Methodology grouped into the three categories developed by the Task Force on POPs.

Pesticides	Industrial chemicals	By-products
Aldrin	Hexabromobiphenyl	Dioxins
Chlordane	PCBs	Furans
Chlordecone	Pentachlorophenol	Hexachlorobenzene
DDT	SCCP	PAHs
Dieldrin	Hexachlorobenzene	
Endrin		
Hexachlorobenzene		
Heptachlor		
Lindane		
Mirex		
Toxaphene		

Note: During the assessments, hexachlorobenzene was listed in all three categories.

sources and developed control options for different kinds of such sources. These options consisted of combinations of best available techniques and emission limit values. While some states preferred specific technology-based controls on each source as the most effective way to reduce emissions, other argued for an emission limit system, which would set limits for overall emission levels within a given geographical area, allowing flexibility in emission amounts from individual sources as long as the overall goal was achieved. On all the main policy issues, the preparatory working group helped focus CLRTAP regulatory efforts and set important precedents for the number of chemicals covered by the global work on POPs developing at the same time. However, final political decisions were deferred to the protocol negotiations.

On the design of the track 2 mechanism, the preparatory working group discussed two main proposals. Some countries, including Canada and the United States, argued in favor of a mechanism that set detailed specifications for substance data, including numerical values for the scientific assessment criteria, together with comprehensive stipulations for how the substance assessments should be performed and by whom. This approach, seen to provide clarity and conformity in data requirements and assessment procedures, was also preferred by the chemical industry, for which the mechanism for future evaluation of additional substances was particularly important. Most substances identified for initial controls were either phased out or in only limited production and use in the CLRTAP region. However, substances with much higher commercial value were likely to come up for future evaluation, raising the economic and political stakes.

Viewing this approach as too rigid, a coalition of northern European states, led by the EU, argued that having detailed numerical criteria specified in the protocol could be counterproductive. They argued that scientific knowledge of POPs advances continuously, and any criteria specified in detail could become outdated, binding the parties to obsolete requirements. By having less detailed criteria, each future evaluation could be based on the latest available scientific information and criteria developments. Also, European countries wanted to avoid specified numerical criteria because they believed that it would send an inaccurate message that there is a clear line between hazardous and harmless substances, instead preferring to give greater prominence to precaution. In the end, the preparatory working group could not reach consensus on this issue, and the two main policy options were carried over to the protocol negotiations.

Protocol Negotiations

The CLRTAP parties by 1996 agreed that the preparatory working group had developed sufficiently detailed policy alternatives to begin protocol negotiations. Five negotiating meetings were held in 1997 and 1998.[4] Three main coalitions that had begun to form during the assessments dominated the negotiations: the EU, supported by Norway and Switzerland; the United States, often together with Canada and typically closest to the opinions of the chemical industry; and Russia, frequently backed by Ukraine (H. Selin 2000). Much of the substantive negotiating took place among these three coalitions, with other states largely passive. In general, the EU and its supporters backed by NGOs sought the most stringent regulations in the face of opposition from the other two blocs, foreshadowing dynamics that were also present during the negotiations of the Stockholm Convention.

Several IGOs and NGOs observed the negotiations (H. Selin 2000). UNEP Chemicals, preparing to start global negotiations on POPs in 1998, was the main IGO in attendance. Many of the same industry organizations that followed the assessments dominated the NGO group. Greenpeace, the single environmental NGO to follow the negotiations on site (others paid more attention to the global work), attended only one meeting. However, reflecting the importance that many Arctic indigenous peoples gave to the POPs issue, representatives from the ICC attended a few of the later meetings (Fenge 2003). ICC representatives lobbied successfully for the Arctic region to be recognized as particularly sensitive to POPs. As a result of their lobbying, the preamble of the protocol gives special recognition to the exposed situation of the Arctic environment and human populations (N. Selin and H. Selin 2008).

The extensive preparatory work had clarified three main sets of issues in need of political solutions during the negotiations: the initial list of regulated substances, the format for their controls, and the mechanism for future evaluation of additional substances (H. Selin 2000).

Initial List of Substances

The CLRTAP assessments had identified eighteen candidate POPs for the initial list. Among the fifteen pesticides and industrial chemicals, seven were relatively uncontroversial: aldrin, chlordane, dieldrin, endrin, mirex, hexabromobiphenyl, and toxaphene were no longer in production or use in the CLRTAP region by the late 1990s and did not require any new domestic restrictions. The other eight pesticides and industrial chemicals, however, were still in production and use. There was general acceptance that chlordecone, hexachlorobenzene, DDT, and PCBs should be included in the protocol, and the negotiations focused on whether they should be banned or if certain uses should be permitted. In contrast, there were disagreements about the inclusion of lindane, heptachlor, pentachlorophenol (PCP), and SCCP. In addition, countries agreed that the four by-products should be covered by the protocol.

At the first negotiation meeting, countries agreed to ban chlordecone, an insecticide chemically similar to mirex, which was also banned under CLRTAP. New scientific data demonstrated that chlordecone had more severe bioaccumulation and toxicity potential than previously thought. Furthermore, its use had declined steadily since the late 1970s. With few uses in the region by the time of negotiations, there were no major objections to a phaseout. Negotiations also addressed hexachlorobenzene, which was sometimes used as an industrial chemical to make fireworks, ammunition, and synthetic rubber, but its main application was as a pesticide for seed treatment. Russia, the Ukraine, and the United States were against a ban on hexachlorobenzene. Because of their opposition, exemptions were granted during the negotiations for both the production and use of hexachlorobenzene for specified seed treatment purposes.

A majority of countries were strongly in favor of a CLRTAP ban on DDT, but Italy, the Russian Federation, and the United States sought exemptions. Italy desired—and gained—an explicit exemption for the use of DDT in the manufacturing process for the pesticide dicofol. The Russian Federation wanted to continue to produce and use DDT domestically for use against malaria carrying mosquitoes. The United States, referring to the WHO policy recommendation that DDT in some cases was the best available means to fight malaria (see chapter 3), wanted an exemption for the export of DDT for malaria treatment (having banned its use domestically in the wake of *Silent Spring*). As a result of the positions by the Russian Federation and the United States, DDT exemptions were granted for public health protection as a component of an integrated pest management strategy against malaria and encephalitis.

During the assessments, Russia maintained that the manufacturing and use of PCBs had ceased during the Soviet era. However, at the third negotiations session in 1997, Russia suddenly claimed that it needed exemptions for the use of PCBs in transformers and suggested that domestic production may still exist. The Russian announcement caused great surprise and sparked several activities. One was an effort to try to formulate treaty text that would be acceptable to both Russia and those who wanted a complete ban on PCBs. As a compromise, a five-year extension on PCB phaseout was granted to countries with economies in transition. In addition, multilateral and bilateral processes were initiated outside CLRTAP to explore ways in which European and North American countries could assist Russia in stopping the use of PCBs and managing stockpiles in an environmentally sound manner.

The pesticide lindane attracted much attention during the negotiations. Canada and the United States questioned data on its bioaccumulation. However, western European states argued that it warranted inclusion on the basis of the precautionary principle because of its general chemical and biological characteristics (Sweden 1997a). Furthermore, addressing only lindane could lead to increased use of technical hexachlorocyclo-hexane (HCH), a simpler (and more toxic) version of lindane that had already been largely phased out. As a result, HCH was included in the POPs protocol in two categories: technical HCH and lindane. No exemptions on technical HCH were given, and lindane was subject only to restricted use. Thus, as most Europeans desired, lindane was subject to controls, but permitted exemptions covered almost all known uses in North America, including against head lice, the main exemption sought by the United States. Thus, few new lindane restrictions were introduced by the protocol.

A prominent example of CLRTAP actions being influenced by the global POPs process related to the handling of heptachlor. In 1995, the UNEP Governing Council listed heptachlor, the only UNEP POP that was not on the CLRTAP list at the time, as one of twelve POPs warranting global action. Under the modified task force methodology screening, heptachlor did not meet the criteria of a POP; however, when heptachlor was identified by UNEP, many European countries quickly suggested that it should be included on the CLRTAP list based on a precautionary approach. Yet there was some uncertainty about heptachlor's propensity for long-range transport. In addition, the United States insisted that it needed heptachlor to combat fire ants in industrial electrical junction boxes. In the end, heptachlor was included in the protocol with a U.S. exemption for its use, with a commitment to reevaluate this use at a later date.

While some European countries referred to the precautionary principle to argue in favor of regulating the pesticide PCP, the United States and Canada believed that the data on bioaccumulation was unclear (United States 1997a, 1997b). On the industrial chemical SCCP, states disagreed whether it met the long-range transport criteria. While Canada and the United States argued that there was a lack of conclusive scientific evidence, the Netherlands and Sweden were among the strongest advocates for its inclusion, again referring to the precautionary principle (Sweden 1996, 1997b). In the end, PCP and SCCP were deleted from the initial list because of a lack of consensus.[5] As a result, the CLRTAP POPs Protocol initially covered sixteen chemicals (see table 6.3). Nine or ten of these are also regulated under the Rotterdam Convention, depending on whether HCH and lindane are considered one or two substances.[6]

Table 6.3
Substances initially included in the CLRTAP POPs Protocol

Annex	Pesticides	Industrial chemicals	By-products
I	Aldrin Chlordane Chlordecone DDT Dieldrin Endrin Heptachlor Hexachlorobenzene Mirex Toxaphene	Hexabromobiphenyl PCBs Hexachlorobenzene	
II	HCH/Lindane DDT	PCBs	
III			Dioxins Furans Hexachlorobenzene PAHs

Note: The substances are grouped according to category and the annex in which they are listed. DDT and PCBs are listed in both annex I and annex II; hexachlorobenzene is listed in both annex I and annex III.

Format for Controls

All countries agreed to ban uses of the POPs scheduled for elimination. Only the United States objected to banning production as well, because it wished to retain the possibility of producing POPs banned by the CLRTAP POPs Protocol for export to countries outside the UNECE region. In contrast, most European countries believed that an export ban was desirable, if only because its absence could lead to accusations of a double standard: banned POPs were considered too dangerous for domestic use but not too dangerous for export to nonparties (mostly developing countries). Exported POPs could also have negative environmental and human health effects within the CLRTAP region because of long-range transport. The United States and Canada, however, argued that trade restrictions were inappropriate in regional environmental agreements, linking the development and operation of the CLRTAP POPs Protocol with the Rotterdam and the Stockholm conventions on issues relating to the trade in commercial chemicals.

Not until the very final stage of the negotiations did the two North American and European coalitions reach a compromise. In the end, Canada and the United States convinced their European counterparts that import and export controls should not be included in the CLRTAP POPs Protocol because of its regional scope. The EU also believed that it could return to the trade in POPs under the Stockholm Convention, which was then set to be negotiated. In turn, the United States lifted its reservation on introducing production bans. As a result, the pesticides and industrial chemicals that are regulated under the CLRTAP POPs Protocol are grouped into two separate annexes (see table 6.3). Pesticides and industrial chemicals for which parties are required to eliminate both domestic production and use are listed in annex I. Pesticides and industrial chemicals that are subject to use restrictions, but where specific use or production exemptions are given, are listed in annex II.

Many European countries favored restrictions on the transboundary transport of POPs waste, arguing that such trade increased risks for leakages and emissions. Again, the United States and Canada objected to trade restrictions, also arguing that permitting the transport of POPs wastes would help ensure that they were disposed of in both a cost-effective and environmentally sound manner. As a result, parties commit to "endeavour to ensure" that the disposal of regulated POPs is carried out domestically, but transboundary wastes transfers are permitted. The CLRTAP POPs Protocol also stipulates that the terms *waste, disposal,* and *environmentally sound* should be interpreted "in a manner consistent with" the use of those terms under the Basel Convention (see chapter 4). This issue— and how the treaty text was formulated—was somewhat controversial because the United States was not a Basel party and wanted to avoid establishing direct legal commitments between the two treaties.

Negotiations on the by-products focused on the design of different types of controls and whether they should be mandatory or voluntary. The four by-products listed in annex III are regulated through a combination of emission limit values and best available techniques controls specified in detailed technical annexes. Parties should reduce total annual emissions of the four by-products from twelve types of major stationary sources based on a self-selected reference year between 1985 and 1995. Parties are required to set emission limit values for dioxins and furans from each new stationary source (annex IV) and are recommended to apply the best available techniques on each new major stationary source for dioxins, furans, PAHs, and hexachlorobenzene (annex V). Furthermore, parties must apply emission limit values and best available techniques to existing stationary sources if they are technically and economically feasible. Finally, the protocol identifies recommended (rather than mandatory) control measures for reducing emissions from mobile sources (annex VII).

Mechanism for Future Evaluation of Additional Substances

The design of the mechanism for evaluating additional candidate POPs was another key issue during the negotiations with strong linkages to the global POPs work. Carrying over the discussions from the preparatory working group, Canada and the United States, with support from the chemical industry, restated a preference for a mechanism whereby a proposed chemical would have to meet detailed numerical scientific criteria, listed in the protocol, for assessing harmfulness: persistence, toxicity, and bioaccumulation. They argued that this approach would incorporate transparency and consistency by having clearly defined criteria against which to assess all substances. A second coalition of mostly western European states advocated for a more flexible mechanism outside the protocol that could be more easily amended. The proponents of this approach reiterated that embedding detailed criteria in the treaty text could render the mechanism obsolete as knowledge and assessment methodologies advanced.

Countries' positions on the design of the track 2 mechanism were influenced by the fact that they were engaged in precedent-setting work that would have a strong impact on the future direction of international chemicals-screening work and the negotiations and operations of the Stockholm Convention. That is, decisions under CLRTAP would frame the global debate. The first two CLRTAP negotiation meetings clarified the two main policy options (stringent criteria versus flexibility) but made no real progress in reaching an agreement. It was not until the third meeting, when Canada presented a proposal to specify assessment criteria in a separate executive body decision before the signing of the protocol, that the opposing coalitions were able to move forward. As well as offering the predictability sought by Canada and the United States, the proposal allowed amendments to the assessment mechanism to be made by the executive body without having to renegotiate the entire protocol.

Adopted in 1998, Executive Body Decision 1998/2 outlines the procedure for both tightening exemptions for substances on the initial list and assessing substances that are not initially covered by the protocol. In either case, parties must put together a basic risk profile for each substance for which expanded CLRTAP action is proposed. A risk profile should contain data on atmospheric transport, toxicity, persistence, and bioaccumulation, as well as data on production, uses, emissions, environmental levels in areas distant from sources, and environmental degradation. In addition, four kinds of socioeconomic information must be included in the risk profile: (1) alternatives to existing uses and their efficacy, (2) adverse environmental or human health effects associated with alternatives, (3) pollution prevention technology and techniques that may reduce emissions, and (4) costs and benefits of alternatives.

The procedure for evaluating a new substance for addition to the protocol contains several steps. Based on the risk profile, the parties initiate a technical review of the proposed chemicals by a special task force on POPs to determine whether further consideration may be warranted. This review evaluates data on long-range transboundary transport, adverse environmental and human health effects of long-range transboundary atmospheric transport, emission sources, and the technical feasibility and associated effects and costs of available abatement options. If the task force concludes that action may be justified, it submits a report to the Working Group on Strategies and Review and the executive body. If the executive body classifies a proposed chemical as a POP based on the technical review, it can instruct the task force and the Working Group on Strategies and Review to develop proposals for regulatory action under the protocol, to be approved by all parties by consensus.

Implementing the Protocol

In 1998, at the Fourth Environment for Europe Ministerial Conference in Århus, Denmark, thirty-three countries and the EU signed the CLRTAP POPs Protocol. At the meeting, the European Commission and the EU member states issued a declaration stating that they welcomed the protocol, but would in fact have preferred a more stringent agreement. The protocol entered into force in 2003. As of late 2009, twenty-eight countries and the EU had ratified the agreement. Among the major CLRTAP countries, the United States, Russia, and Ukraine have not ratified the protocol. During the implementation of the protocol, the parties have focused on two major sets of issues: organizational developments and expanding POPs regulations. These issues through both governance and actor linkages also connect to decision processes and management activities under other treaties, influencing the effectiveness of the chemicals regime.

Organizational Developments
The CLRTAP Executive Body oversees the implementation of the POPs protocol, and the CLRTAP secretariat acts as a coordination center for the gathering of information on international and domestic implementation measures. The Co-operative Programme for Monitoring and Evaluation of the Long-Range Transmission of Air Pollutants in Europe (EMEP) monitors domestic implementation. To that end, EMEP works on creating standard operating procedures and quality control routines for sampling and chemical analysis (UNECE 1999). EMEP also operates an international measurement program on POPs. In their work, two EMEP centers—the Chemical Coordinating Centre and the Meteorological Synthesizing Centre-East—initially focused on a small group of selected POPs that included lindane, PAHs, PCBs, dioxins, and furans. These efforts also contribute to the global work on POPs, as most of these chemicals are also covered by the Stockholm Convention.

Specifically, the EMEP Chemical Coordinating Centre focuses on developing standard operating procedures for sampling and chemical analysis of POPs. It also works on the operation of an environmental measurement program for POPs, using five major sampling sites: Scandinavia and the Baltic, the northern Atlantic region, continental Europe, the Mediterranean region, and the south Atlantic region. The EMEP Meteorological Synthesizing Centre-East in Russia is responsible for modeling POPs, studying the chemical properties of the selected POPs, and analyzing and summarizing scientific results obtained under international and national programs. The Meteorological Synthesizing Centre-East cooperates with the Meteorological Synthesizing Centre-West in Norway and the experts of the CLRTAP Task Force on Emission Inventories and Projections in the verification of POPs emission data quality.

Executive Body Decision 1998/2 established an Ad Hoc Expert Group on POPs to begin preparing for the assessments of additional chemicals (UNECE, 2000). The expert group met four times between 2000 and 2003.[7] Collaborating with EMEP, it collected assessment data and elaborated on assessment procedures and guidelines to come into effect once the protocol entered into force. In addition, the expert group discussed substance-specific data and developed scientific dossiers on SCCP, endosulfan, and dicofol. After the protocol entered into force, a formal Task Force on POPs was created by the Executive Body (UNECE 2004). The task force, which is open to all CLRTAP parties, assesses technical reviews of assessed chemicals and collaborates with the Working Group on Strategies and Reviews in exploring management option under the guidelines of Executive Body Decision 1998/2.

Expanding POPs Regulations

The Task Force on POPs met six times between 2004 and 2007 to consider several changes to the protocol.[8] Experts from approximately twenty parties—most of them from leading EU member states plus representatives of Canada, the European Commission, Norway, Russia, Switzerland, and the United States—attended the task force meetings together with several observers, most of whom were from the chemicals industry. CLRTAP countries including the United States that are not yet parties to the protocol can partake in the task force discussions but cannot propose additions and do not have voting rights in the executive body. Although task force discussions are formally independent from discussions outside CLRTAP, there are strong linkages to parallel efforts to expand regulations under the Rotterdam and the Stockholm conventions. That is, agreements and disagreements among CLRTAP parties (which are all leading regime participants) on a particular chemical carry over to other forums.

Based on their positions during protocol negotiations, as well as statements in the ministerial declaration issued when the treaty was adopted, the European parties have taken the lead in proposing additional chemicals. Norway was the first party to propose a chemical when it nominated pentabromodiphenyl ether (penta-BDE) in 2003. This was quickly followed by Sweden's proposal to include perfluorooctanesulfonate (PFOS) in 2004. The executive body, based on the technical reviews and recommendations by the task force, agreed that they meet the CLRTAP criteria for a POP at its meeting in 2005 (UNECE, 2006). In addition, the European Commission proposed in 2005 the inclusion of five additional substances: hexachlorobutadiene (HCBD), octabromodiphenyl ether (octa-BDE), pentachlorobenzene (PeCB), polychlorinated naphthalene (PCN), and SCCP. The executive body again recognized all of these chemicals as POPs at its meetings in December 2006 and tasked the Working Group on Strategies and Review to develop abatement proposals (UNECE 2007a, 2007b).

The negotiations on the seven substances classified as POPs by the executive body—together with changes to the controls of chemicals already in the POPs Protocol and stipulations in the technical annexes—have been difficult. Parties express differing opinions on how specific POPs should be regulated and what kind of exemptions should be allowed. In 2008, Norway and the EU furthermore proposed new additions: Norway proposed hexabromocyclododecane, and the European Commission put forward trifluraline, dicofol, pentachlorophenol, and endosulfan. This not only expands the number of new chemicals considered under CLRTAP, but creates further policy-making and management linkages under the chemicals regime. The European Commission and individual European countries push the same chemicals in CLRTAP, the Stockholm Convention, and the Rotterdam Convention. Many of these chemicals, including endosulfan, are also subject to contested negotiations across regional and global policy forums (see chapters 5 and 7).

Conclusion

The 2004 and 2009 AMAP reports concluded that many POPs remain problematic in the Arctic. Assessments in the Baltic Sea region and the Northeast Atlantic show similar results (H. Selin and VanDeveer 2004). Some of these continuing problems are the result of the very slow environmental degradation of many POPs already covered by the CLRTAP POPs Protocol. More worrying are indications that hazardous substances not covered by the protocol can be found at high levels in environmental samples throughout the UNECE region. Clearly there is a need to carefully assess implementation progress on regulated chemicals as well as continue to evaluate and add substances to the protocol. These issues are critical to the effectiveness of both the CLRTAP POPs Protocol and the chemicals regime. However, efforts to amend the protocol have been very slow.

As CLRTAP parties work to expand regulations on POPs, many important data gaps remain. Continued CLRTAP policy making and implementation on POPs would benefit from expanded international efforts to improve data availability, quality, and comparability for both environmental and socioeconomic information. This is also a major goal of SAICM. The CLRTAP work on POPs so far has also been primarily reactive: chemicals have become subject to regulations only after they have been found causing damage to wildlife and humans. Efforts to more actively identify additional POPs before they cause environmental and human health damage require developing more effective methods of assessing both existing and new commercial chemicals. It is also important to improve understanding of when POPs are generated as by-products and to develop better and cheaper technology and techniques for preventing emissions.

The implementation of the CLRTAP POPs Protocol is connected to many other parts of the chemicals regime. CLRTAP waste rules are designed to complement the Basel Convention. CLRTAP and the Basel Convention also overlap with the work of the Basel Convention regional centers in Moscow and Bratislava on managing POPs waste. The Rotterdam Convention relates to the CLRTAP POPs Protocol in those cases where a CLRTAP substance is also on the PIC list or where a chemical is considered for inclusion under either treaty. Yet CLRTAP has the closest ties with the Stockholm Convention because parties to both treaties assess additional substances. Furthermore, the development of regional centers under the Stockholm Convention will create additional linkages with CLRTAP as well as the Basel Convention. Finally, CLRTAP abatement efforts are also dependent on global actions to reduce long-range transport of emissions that originate from outside the region.

7
The Stockholm Convention and Global POPs Controls

This chapter analyzes global policy and management efforts on POPs. The significant PCB problem in Tanzania noted in the chapter 1 is not unique, but countries worldwide are struggling to manage PCBs as well as other POPs. In Nepal, for example, tens of thousands of workers in welding workshops were found to be handling PCB-contaminated oils with little awareness of the human health dangers of PCBs (Nepal 2007). A Cambodian assessment noted, "The exposed employees and workers who work in transformer workshops, warehouses, and power plants are unaware of PCBs hazard. For instance, they have pumped PCBs oil from container to transformer, and pumped used PCBs oil from one broken transformer to another by siphoning by mouth. In addition, employees and workers have often used PCBs oil to paint furniture and to sell to other people for various uses" (Cambodia 2006, 35).

Analyzing the development of global POPs controls, this chapter focuses on the creation and implementation of the Stockholm Convention. The Stockholm Convention aims to protect human health and the environment from POPs by introducing life cycle controls on specific pesticides and industrial chemicals and by setting standards for regulating emissions of by-products. The chapter focuses on several key issues related to the formation of actor coalitions and to policy diffusion that are of importance to the effectiveness of the Stockholm Convention and the chemicals regime. The issues include the expansion of regional POPs controls to the global level, the formation of the POPs Review Committee for the evaluation of additional chemicals, the creation and funding of organizational structures supporting capacity building, and the establishment of mechanisms for monitoring and compliance.

The chapter begins by examining how the POPs issue was picked up by global IGOs and a larger set of countries in the mid-1990s. The next section discusses the importance of the global assessment and awareness-raising efforts that were conducted from 1995 to 1998, and the ways in which those activities were heavily influenced by governance and actor linkages with the CLRTAP POPs work. This is followed by an analysis of the major actor coalitions and policy issues during the negotiations of the Stockholm Convention, which began in 1998 and ended in 2001. Next, the chapter examines efforts to implement the Stockholm Convention, including how these efforts are shaped by linkages with other forums. The chapter ends with a few remarks on major policy and management issues related to the continued implementation of the Stockholm Convention and the effective multilevel management of POPs.

POPs Go Global

The Stockholm Convention has been developed through more than a decade-long process (see table 7.1). Scientific work in the Arctic region and the CLRTAP assessments helped construct POPs as a distinct subissue of chemicals management through close interactions among scientists and

Table 7.1
Chronology of important events in the creation and implementation of the Stockholm Convention.

Time	Event
May 1995	UNEP Governing Council begin assessments of 12 POPs
June 1996	IFCS Ad Hoc Working Group on POPs concluded that the scientific data justified creating a global treaty
June–July 1998	First negotiation session of the Stockholm Convention
May 2001	Stockholm Convention adopted
May 2004	Stockholm Convention enters into force
May 2005	COP1 held
May 2009	COP4 adds nine more substances, for a total of twenty-one regulated chemicals
July 2009	163 states and the EU were parties to the Stockholm Convention

policy makers, and the CLRTAP definition of a POP was largely copied by the IGOs, NGOs, and states involved in the global negotiations. As a result, the CLRTAP framing of the POPs problem as an issue of long-range transport of toxic persistent substances carries through to the present as the dominant global description of POPs hazards, including in the Stockholm Convention (see chapter 6).

Although some POPs have attracted attention for their hazards since at least the 1950s, scientific and political concern has increased significantly since the 1980s regarding the dangers that POPs pose to the environment and human health. POPs were discussed during some preparatory sessions for UNCED, and chapter 19 of agenda 21 stressed the importance of more effective chemicals management, but the movement toward global POPs regulations did not begin in earnest until the mid-1990s. In 1995, the UNEP Governing Council called for global assessments of twelve specific POPs—sometimes referred to as "the dirty dozen" (see table 7.2) (UNEP Governing Council 1995). Importantly, it was only after CLRTAP's assessment work had produced a state-of-knowledge report and the CLRTAP parties began considering creating a POPs protocol that major IGOs and NGOs took a more sustained interest in the POPs issue.

The UNEP-initiated global POPs assessments were coordinated by a set of IGOs already working on a range of chemicals management issues (Downie 2003). Based on the UNEP Governing Council decision in 1995, the IOMC, IPCS, and IFCS coordinated the assessments. The main assessment work was conducted by the IFCS Ad Hoc Working Group on POPs, whose members included government-appointed experts as well as

Table 7.2
POPs selected by the UNEP Governing Council

Pesticides	Industrial chemicals	By-products
Aldrin	Hexachlorobenzene	Dioxins
Chlordane	PCBs	Furans
DDT		
Dieldrin		
Endrin		
Heptachlor		
Mirex		
Toxaphene		

representatives from NGOs and industry. The working group focused on compiling and assessing information on the chemistry, sources, toxicity, environmental dispersion, and socioeconomic impacts of the dirty dozen. To that end, the then nearly complete CLRTAP assessments were extensively used as a data source.

The use of CLRTAP information in the global process allowed the assessments to be conducted relatively quickly because key data had already been collected and vetted. In fact, the UNEP Governing Council asked that the global assessments take into account information collected under the CLRTAP assessment process as a way to ensure their scientific credibility. Of use were the CLRTAP POPs assessments of the physical nature of the problem and identification of policy options. For the former, information about use areas of substances, emission sources, transport patterns, and environmental and human health effects was used The CLRTAP work on policy options helped identify important activities that needed to be covered by the global agreement and provided suggestions on how controls and procedures could be designed (H. Selin and Eckley 2003).

Furthermore, the Conference to Adopt a Global Programme of Action for the Protection of the Marine Environment from Land-Based Activities, held in Washington, D.C., in 1995, adopted a declaration supporting the global assessments. The declaration also recommended the development of a global legally binding instrument to control POPs. The IFCS Ad Hoc Working Group on POPs presented its findings at an expert meeting in 1996. These findings were largely consistent with the CLRTAP assessments, defining the POPs issue as one of long-range transport and deposition, with POPs having adverse environmental and human health effects both near and distant from their emission sources. The IFCS Ad Hoc Working Group on POPs concluded that the scientific data justified the creation of a global POPs treaty. In response, the UNEP Governing Council in 1997 requested that UNEP begin treaty negotiations.

The same states that pushed for CLRTAP action on POPs were strong advocates of global regulations. In fact, CLRTAP leader states early on advocated for a global POPs agreement, but it was only after the CLRTAP scientific work had progressed that global action became possible. To this end, the CLRTAP leader states used their regional activities to scale up policy efforts and initiate work on a global treaty. As in CLRTAP, Canada and Sweden played important intellectual and material leadership roles.

They were highly active in both UNEP and the IFCS, sponsoring meetings, preparing assessment reports, and developing policy options. Additionally, the first convention negotiation session took place in Montreal, sponsored by Canada, and the chair of the negotiations was a Canadian. Sweden early on volunteered to hold the diplomatic conference adopting the treaty in Stockholm, naming the treaty the Stockholm Convention.

The global assessments also included industrialized countries outside the CLRTAP region (e.g., Australia, New Zealand, and Japan), as well as the world's developing countries. In particular, the addition of the large number of developing countries affected the global assessments. The UNEP Governing Council in 1995 specifically stated that the global assessments should take into account the circumstances of developing countries and countries with economies in transition. To this end, a large part of the global assessment process aimed at building awareness and capacity among states not involved in the CLRTAP process. UNEP and the IFCS held eight regional awareness raising workshops in 1997 and 1998, including in Bangkok, Buenos Aires, St. Petersburg, and Bamako. At these meetings, government representatives from countries all over the world presented domestic case studies to other national delegates (H. Selin and Eckley 2003).

The regional workshops made the POPs issue salient to a larger group of countries outside the CLRTAP region, in many cases prompting them to assess their domestic status of POPs for the first time (N. Selin 2006). For most developing countries, their domestic situation regarding POPs in the 1990s was fundamentally different than it was for the CLRTAP countries. Industrialized countries had often phased out much production and use of the commercial POPs among the dirty dozen even before the CLRTAP negotiations began, if they have been used at all. For example, in Sweden, several POPs pesticides were never approved, and the country banned endrin in 1966 and aldrin and dieldrin in 1970 (Swedish Chemicals Inspectorate 2006). In contrast, Tanzania imported over 110,000 liters of dieldrin in 1989, and domestic use of many POPs pesticides continued well into the 1990s (Tanzania 2005).

The CLRTAP assessments, focusing largely on problems associated with long-range transport in the Northern Hemisphere, did not address many issues important for developing countries. The inclusion of a large group of developing countries in the negotiation of the Stockholm Convention added a much stronger focus on local-level management and capacity issues than under CLRTAP (although these issues had also been discussed in CLRTAP, specifically with regard to the situation of eastern European countries). Many developing countries viewed POPs much more as a local management and contamination problem that because of extensive chemicals import also had a strong trade component. Technical and financial assistance to deal with pesticides and chemicals became a top priority for many developing countries during the negotiations, as in negotiations of the Basel and Rotterdam conventions (see chapters 4 and 5).

Negotiating the Stockholm Convention

Negotiations on the Stockholm Convention started in June 1998, three months after the Rotterdam Convention had been finalized and the same month that the CLRTAP POPs Protocol was signed. Many of the same people who had participated in the negotiations of these two earlier treaties were also negotiating the Stockholm Convention, in effect linking the three treaties. Several global IGOs played key roles during the negotiations, which were organized under UNEP auspices, with UNEP staff providing important administrative and other kinds of support services during and between negotiating sessions. In addition, WHO, ILO, GEF, and UNITAR brought their respective areas of expertise to the meetings. In general, these organizations were strong supporters of the effort to create a global POPs treaty, and their representatives participated actively in the negotiating sessions.

Most of the world's countries attended the convention negotiations, which were relatively harmonious, especially during the first couple of meetings. The CLRTAP countries had found common ground during the protocol negotiations and were now seeking to elevate the regional controls to the global level. Countries outside the CLRTAP region shared many of the same concerns, as the IFCS-led assessment work and workshops organized between 1995 and 1998 had established a platform for productive treaty negotiations based on a shared acknowledgment of the POPs threat. In addition, all countries benefited from positive linkages in the form of discussions on several procedural and legal issues during the negotiations of the Rotterdam Convention, aiding the development of the

Stockholm Convention. Nevertheless, there were also important disagreements among countries that had to be overcome as the final stages of the negotiations became more contentious.

Leading industry organizations and firms followed the development of the Stockholm Convention more closely than the CLRTAP POPs work. The ICCA represented leading industry interests, working closely with ACC, CEFIC, and other regional associations of firms. Most industry participants supported efforts to regulate the twelve chemicals identified by the UNEP Governing Council. As the global negotiations started, the CLRTAP POPs Protocol had just been concluded, introducing comprehensive regional regulations on the ten pesticides and industrial chemicals that were among the dirty dozen. Dominating industry interests in Europe and North America—like their home governments—backed the expansion of these regulations to the global level so that their competitors and chemicals producers across the world would operate under similar restrictions.

By the late 1990s, most industrialized countries, together with a growing number of developing countries, had banned the production and use of the ten commercial POPs under negotiation or had restricted their application. As a result, the commercial value of these substances was relatively low to the major chemical companies by the time the treaty negotiations began. Leading industry actors in Europe, North America, and elsewhere therefore stood to lose little with strong global regulations and could even perhaps gain some business opportunities as remaining users looked to switch to alternatives. In addition, the ICCA and other industry representatives actively followed the development and design of the mechanism for the evaluation of additional chemicals after the entry into force of the Stockholm Convention, as countries were expected to propose chemicals with higher market value in the future.

NGO participation was also higher during the global negotiations compared with CLRTAP. Many NGOs observed the negotiations through the International POPs Elimination Network, a network of over 400 advocacy groups from all over the world that was founded in 1998 to support the development of strong POPs controls. The network was officially launched at the first negotiating session of the Stockholm Convention. In addition, representatives of major environmental NGOs such as Greenpeace and the WWF attended the negotiations. NGOs issued statement papers and made frequent interventions during plenary meetings as they lobbied in support of stringent controls of POPs. In addition, Arctic indigenous peoples groups were active participants from the beginning of the negotiations. Among these groups, the ICC played a particularly high-profile role, drawing on experiences from the CLRTAP negotiations.

The ICC represents the approximately 150,000 Inuit living in Alaska, Canada, Greenland, and Chukotka of (Russia).[1] Arctic indigenous groups have long supported the expansion of international law as a means for advancing their political agenda of promoting sustainable land management and human health protection. They have done so in the context of seeking increased self-determination within their own states, but typically stopping short of advocating for secession (Koivurova and Heinämäki 2006). In the Arctic, indigenous groups have been particularly active in the Arctic Council, including on hazardous substances (H. Selin and N. Selin 2008). However, these groups, together with the eight Arctic states, have also been vocal advocates for legal developments outside Arctic forums to protect the region from transboundary hazards (Fenge 2003).[2]

Several Arctic indigenous groups participated in an NGO-organized forum prior to the beginning of the negotiations (Watt-Cloutier 2003). At the first negotiating session, Sheila Watt-Cloutier, then Canadian president of the ICC, stressed the public health threat from POPs contamination in Arctic food webs. Watt-Cloutier also presented UNEP's executive director, Klaus Töpfer, with an Inuit soapstone carving of a mother and child during the second negotiating session. John Buccini, the chair of the negotiations, carried this statue to every subsequent meeting as a moral symbol of what was at stake. The concerns voiced by the Arctic indigenous groups are also reflected in the preamble of Stockholm Convention. Similar to the text in the CLRTAP POPs Protocol, the Stockholm Convention recognizes that Arctic ecosystems and indigenous communities are particularly at risk because of the biomagnification of POPs and the contamination of traditional foods.

During the three years it took to finalize the Stockholm Convention, IGO staff, country officials, and NGO representatives met during five week-long negotiation sessions, in two week-long meetings of the Criteria Expert Group tasked with developing scientific criteria and procedures for adding POPs to the treaty after its entry into force, and during one meeting focusing exclusively on financial assistance.[3] Three sets of issues were particularly controversial during the negotiations: the design of control measures for the twelve POPs, including exemptions to allow some continued uses; the development of science-based criteria and a process for evaluating and adding POPs to the treaty in the future; and issues relating to technical and financial assistance and capacity building.

The Initial Substance List

One of the most striking ways in which the CLRTAP assessments influenced the global POPs work was in the selection of substances for global agreement (H. Selin and Eckley 2003, N. Selin 2006). The list of twelve chemicals identified by the UNEP Governing Council in 1995 was based on the then ongoing assessment scheme in CLRTAP (see chapter 6). Under the CLRTAP Protocol, the selection of substances for action was a significant area of disagreement, and the initial list of regulated substances was not finalized until the end of the protocol negotiations. In contrast, the selection of the global list was one of the least controversial aspects during the development of the Stockholm Convention. In fact, the list of the dirty dozen remained unchanged and largely unchallenged throughout the global assessments and treaty negotiations from 1995 to 2001.

At the UNEP Governing Council meeting in 1995, a tentative CLRTAP substance list was introduced and accepted with few comments from countries not involved in the CLRTAP assessments. The CLRTAP POPs Protocol eventually addressed four additional substances, but the list of chemicals that the UNEP Governing Council considered resembled the CLRTAP list as it existed in 1995. However, the UNEP and CLRTAP lists were not identical, and regulatory preferences also varied among states in the CLRTAP negotiations. Specifically, there was some confusion over the UNEP inclusion of heptachlor, which at the time was not on the CLRTAP list (H. Selin and Eckley 2003). Heptachlor, however, remained on the UNEP list and was added to the CLRTAP one. As a result, all twelve UNEP POPs were covered by CLRTAP. In addition, eight of the ten commercial Stockholm Convention POPs are also regulated under the Rotterdam Convention.[4]

The Stockholm Convention divides POPs in three separate annexes (see table 7.3). Annex A originally contained seven pesticides (aldrin, chlordane, dieldrin, endrin, heptachlor, mirex, and toxaphene) and two industrial chemicals (hexachlorobenzene and PCBs). The use and pro-

Table 7.3
Substances originally covered under the Stockholm Convention

Annex A	Annex B	Annex C
Aldrin	DDT	Dioxins
Chlordane		Furans
Dieldrin		Hexachlorobenzene
Endrin		PCBs
Heptachlor		
Hexachlorobenzene		
Mirex		
Toxaphene		
PCBs		

duction of annex A chemicals are in general prohibited. Parties may, however, apply for time-limited, country-specific exemptions for annex A substances.[5] Annex B lists POPs that are subject to identified production and use exemptions applying to all parties. Only DDT was initially listed in this annex B. Finally, annex C lists by-products. Four by-products (dioxins, furans, hexachlorobenzene, and PCBs) were originally listed in annex C (hexachlorobenzene and PCBs are regulated as both commercial substances and by-products). Parties should apply best available techniques and best environmental practices for minimizing emissions of by-products.

Similar to their actions during the negotiations of the Rotterdam Convention, developing countries argued for the inclusion of the principle of common but differentiated responsibilities (see also chapter 5). They also believed that this principle should guide the design of specific commitments on individual POPs (Bankes 2003). In contrast to the negotiations of the Rotterdam Convention, where this effort was rejected by industrialized countries that did not want to establish any principal difference among countries' responsibilities on trade-related issues under that treaty, the preamble of the Stockholm Convention explicitly recognizes "the respective capabilities of developed and developing countries, as well as the common but differentiated responsibilities of States as set forth in Principle 7 of the Rio Declaration on Environment and Development." This represented a small victory for developing countries with respect to the management of POPs.

In addition to insisting that the principle of common but differentiated responsibilities guide actions under the Stockholm Convention, a coalition of developing countries, with the support of major NGOs like the International POPs Elimination Network, Greenpeace, the WWF, and the ICC, stressed the importance of the polluter-pays principle for establishing responsibility. Specifically, coalition members argued that POPs-exporting countries should bear a greater burden with respect to their controls (Earth Negotiations Bulletin 1998c, 2000b). Although the preamble of the Stockholm Convention reaffirms the approach "that the polluter should, in principle, bear the cost of pollution," the treaty does not go so far as to give greater legal responsibility to exporting parties (often industrialized countries). Instead, it merely embodies the broader principle of common but differentiated responsibilities.

A coalition of the EU and environmental NGOs pushed for treaty language mandating a phaseout of all POPs toward their elimination, referring to the precautionary principle (Earth Negotiations Bulletin 1999d). Several countries, including the United States, Canada, Australia, New Zealand, Japan, Russia, South Korea, and Thailand, however, argued that a complete elimination was not technically feasible, particularly for by-products (Earth Negotiations Bulletin 1999d, 2000b, 2000c). As a result, the Stockholm Convention merely requires parties to reduce releases of the by-products listed in annex C "with the goal of their continuing minimization and, where feasible, ultimate elimination" (article 5). For commercial POPs, the goal of the Stockholm Convention is to "eliminate releases from intentional production and use" (article 3). In addition, parties have an obligation to ensure that no new POPs become commercially available.

Related to the discussion of what should be the ultimate goal of the Stockholm Convention, negotiators focused attention on specific regulatory measures on the production, use, import, export, and disposal of the ten commercial POPs. While there was broad consensus on the inclusion of all these substances, opinions diverged on for which of them production and use should be completely banned and for what activities and for whom possible exemptions should be granted. In general, countries that argued for an ultimate goal of elimination were the strongest supporters for limiting exemptions on individual POPs covered by the Stockholm Convention. Several other countries, however, argued for provisions that clearly allowed the continued use of some POPs. As a result, the Stockholm Convention allows both for general and country-specific use exemptions.

The Stockholm Convention permits the use of DDT for disease vector control against malaria mosquitoes. To this end, DDT controls are consistent with the aim of the Roll Back Malaria Partnership operated by the WHO, UNICEF, UNDP, and the World Bank (see chapter 3). The WHO estimates that more than 1 million people die from malaria annually.[6] Malaria disproportionately affects Africa—90 percent of all malaria deaths occur on the continent. Other studies that take into account indirect effects of the disease (such as malaria-induced anemia, maternal pathology, and hypoglycemia) estimate that as many as 3 million people may die annually from malaria in Africa alone (Breman, Alilio, and Mills 2004). The high rate of malaria in Africa also has significant economic effects, possibly causing annual economic growth reduction as high as 1.3 percent in the worst affected countries (Gallup and Sachs 2000).

The short-term goal of the Roll Back Malaria Partnership is not eradicating malaria but achieving more effective malaria control (Yamey 2001). The first campaign target was to halve malaria-associated deaths by 2010. Similarly, one of the Millennium Development Goals adopted by the UN General Assembly in 2000 set the goal of having halted and begun to reverse the incidence of malaria by 2015. Indoor spraying of DDT can be highly effective for malaria vector control (Roberts, Manguin, and Mouchet 2000; Rogan and Chen 2005). In the aftermath of the publication of *Silent Spring* and other environmental studies in the 1960s, many industrial countries, IGOs, and NGOs, however, began to support stringent DDT restrictions and phaseouts. During the first negotiating session of the Stockholm Convention, the WHO acknowledged the need for restricted use of DDT for malaria treatment but opposed all agricultural use of the pesticide (Earth Negotiations Bulletin 1998c).

The WHO Expert Committee on Malaria in 2000 recommended the use of DDT "only in well defined, high or special risk situations" for malaria control, but also noted that DDT was being phased out "because of its previous widespread use in the environment, and the resulting political and economic pressure" (WHO 2000). During the convention negotiations, several African countries argued that human health issues related to vector-borne diseases like malaria had to be taken into consideration when designing DDT controls (Earth Negotiations Bulletin 1998c). However, some developing countries announced that they had recently stopped all DDT use, instead emphasizing other means of malaria abatement. Collectively they called for the development of more cost-efficient alternatives. In addition, some NGOs including the WWF had reversed their previous staunch opposition to all DDT application (Earth Negotiations Bulletin 1999d).

DDT and all other substances listed in annexes A and B can be exported and imported only for exempted use or environmentally sound disposal. All trade in commercial chemicals should be in compliance with PIC regulations and requirements under the Rotterdam Convention. The Stockholm Convention also includes provisions for identifying stockpiles, articles in use, and wastes of annex A and B substances. The treaty requires that these be managed and disposed of in an environmentally sound manner. POPs waste may not be transported across international boundaries without taking into account international rules, standards, and guidelines. In this respect, the Stockholm Convention is designed to complement the Basel Convention, and representatives of the Basel Convention secretariat also attended the negotiations and advised on how the two treaties could be linked to capture synergies.

The Mechanism for Future Evaluation of Additional Substances

The Stockholm Convention recognizes that controlling the dirty dozen is only a first step. Similar to CLRTAP, a major issue in which scientific information and political decision making interacted during the negotiations was in the development of a mechanism for adding further substances. Following CLRTAP, the UNEP Governing Council requested in 1997 that the first negotiating session establish an expert group for the development of science-based criteria and procedures for assessing additional substances. The subsequent Criteria Expert Group met twice, once before and once after the second negotiation sessions, drafting a proposal for an assessment procedure that was presented to all countries at the third negotiating session.

As the Criteria Expert Group attempted to establish a set of science-based criteria and an associated procedure for identifying additional candidate POPs for regulatory action, members discussed specific criteria for persistence, bioaccumulation, and toxicity that a substance had to meet in order to qualify as a POP. In other words, the mechanism for adding substances in both the CLRTAP POPs Protocol and the Stockholm Convention is based on specified scientific criteria, consisting of threshold values of bioaccumulation and persistence, combined with a risk characterization demonstrating that a proposed chemical is of international concern. There is, however, no clear, objective scientific threshold that separates a POP from a non-POP, and the selection criteria under both POPs agreements are set based on a combination of science and policy considerations (Rodan et al. 1999).

While the process of establishing evaluation criteria in the Criteria Expert Group was independent from other activities, it is clear that the CLRTAP work in this area played an important precedent-setting role. The two meetings of the Criteria Expert Group were generally constructive, and members finished their work earlier than expected. It is highly likely that the work of the group would have taken much more time if the group members had been forced to start from scratch (N. Eckley 2000). Instead, the CLRTAP assessment criteria were widely accepted because many parties believed that the process had been scientifically rigorous and that the criteria represented a best estimate of where the global negotiations were likely to end up. In addition, the work by the Criteria Expert Group was steered by leading CLRTAP countries that had a strong interest in ensuring that the regional and global criteria were compatible.

As a result, the Criteria Expert Group identified a set of criteria that looked very much like the CLRTAP values, and by the end of the convention negotiations, the global and regional criteria numbers for bioaccumulation and persistence were identical. As such, the work by the group may be viewed as unnecessary given the extensive CLRTAP assessments and the fact that there were large overlaps in membership between CLRTAP and the group. Yet it was critical for the smooth running of the global negotiations that the CLRTAP mechanism was not simply copied without deliberation among both CLRTAP and non-CLRTAP parties. All participants needed to gain a basic understanding and achieve consensus on the choice of criteria, cut-off values, and assessment procedure in order to ensure broad acceptance of the treaty and the procedure (N. Selin 2006).

A major controversy surrounding the design of the assessment mechanism concerned the role of precaution. This debate in many ways mirrored the one under CLRTAP. Based on the fact that there is no simple way to make an objective distinction between a POP and a non-POP, a coalition of European countries, supported by several developing countries and NGOs, believed that an explicit reference to the precautionary principle should be included in the article of the treaty outlining principles and procedures for the assessment mechanism. Another coalition that included the United States, Australia, Canada, Russia, and Japan, however, opposed this. They argued that a more general reference to a "precautionary approach" in the preamble was adequate to guide decision making (Earth Negotiations Bulletin 1999d, 2000b).

In its opposition to the specific inclusion of the precautionary principle in a treaty article, the United States argued that there was no internationally agreed-on definition of the principle, which made its inclusion inappropriate. In addition, countries like the United States and Australia saw a conflict between committing to a science-based assessment mechanism and stressing the importance of precaution in the same article (Earth Negotiations Bulletin 2000c). In contrast, the EU and other supporters of the precautionary principle argued that the principle was adequately defined and there was no contradiction between science-based decision making and precaution. They believed that the inclusion and application of the precautionary principle was critical for guiding decision making on specific substances, as they had also argued for under CLRTAP (Earth Negotiations Bulletin 2000b, 2000c).

In the end, a compromise was reached between the two coalitions where the preamble acknowledges "that precaution underlies the concerns of all Parties and is embedded" within all regulations, requirements, and activities under the treaty. Article 1, outlining the objective of the Stockholm Convention to protect human health and the environment from POPs, recognizes "the precautionary approach as set forth in Principle 15 of the Rio Declaration on Environment and Development." In addition, the importance of precaution is acknowledged in article 8 outlining the review procedure that the conference of the parties shall decide, "in a precautionary manner," if a new chemical warrants controls. This way the EU gained the inclusion of several references in the treaty to the general importance of precaution, but the United States blocked specific mention of the precautionary principle (Bankes 2003).

The development of the review process for additional chemicals also mirrored discussions under the Rotterdam Convention regarding the composition and role of the expert group. The review process consists of five steps. In the first step, any party can submit a proposal to regulate a new chemical based on the information requirement specified in annex D.[7] This proposal is forwarded by the secretariat to the POPs Review Committee, which in step 2 applies the screening criteria in annex D to examine the proposal. The review committee consists of thirty-one government-designated experts appointed by the COP.[8] If the committee deems that the initial screening criteria have been met, the secretariat invites all parties, as well as observers, to provide comments in accordance with the information specified in annex E.[9] In particular, the United States argued that it was critical to allow observers, including industry, to be involved in the review process (Earth Negotiations Bulletin 2000b).

All comments based on the criteria listed in annex E are passed on to the review committee, which in step 3 reviews that information and uses it to develop a substance risk profile. If the review committee decides on the basis of the risk profile that the review should proceed, the secretariat invites all parties and observers to provide additional technical and socioeconomic information, as outlined in annex F, which is forwarded to the review committee.[10] Based on all the available information, the review committee in step 4 formulates a comprehensive risk management evaluation that it submits to the next COP, which in step five takes the final decision by consensus on whether to list the proposed chemical in any of annexes A, B, or C. The COP also specifies related control measures if it is decided that the chemical should be regulated under the Stockholm Convention.

Funding and Capacity-Building Issues

Issues of funding and technical assistance were another key area where developing countries and countries with economies in transition stressed the importance of common but differentiated responsibilities. This discussion was closely related to the development of the national implementation plans that each party is required to complete, outlining a domestic strategy for meeting all obligations under the treaty (Downie 2003). Although both industrial and developing countries acknowledged the importance of financial and technical assistance, they expressed different views on appropriate levels and delivery mechanisms. As such, the conflict between industrialized countries and developing countries over issues of resource allocation and the transfer of financial and technical assistance evident in many other international forums was a critical issue also during the negotiations of the Stockholm Convention.

Many developing countries argued strongly for additional and mandatory assistance from industrialized countries to help them meet their commitments and move toward the use of more environmentally safe alternatives. To this end, they called for the establishment of a new fund organized under the Stockholm Convention. Similar to their earlier demands under the Rotterdam Convention, developing countries argued that such a fund should be modeled after the Montreal Protocol Fund (Earth Negotiations Bulletin 1999c, 1999d). It would provide assistance for a host of POPs projects, with the COPs assessing needs for replenishing the fund. All major industrialized countries, however, rejected this idea, as they had during the Rotterdam Convention negotiations. Again, they preferred to work with existing IGOs, including the GEF (Earth Negotiations Bulletin 1999c, 1999d). Several donor countries also announced voluntary grants during the convention negotiations (Earth Negotiations Bulletin 2000b, 2000c).

In an effort to find a compromise between the opposing coalitions, Canada presented the idea of a capacity assistance network. The idea was that such a network would act as a broker and coordinate available resources with demands for POPs management activities, but without requiring mandatory contributions from donor countries (Earth Negotiations Bulletin 2000b). In addition, representatives of the GEF announced at the fifth negotiating session that the GEF Council had agreed that if the organization were to become the designated financial mechanism for global POPs abatement, GEF would expand its earlier chemical work to POPs issues more broadly. If so, additional financial resources would be made available (Earth Negotiations Bulletin 2000c). In that case, GEF could either operate independently or become a central member of a broader capacity assistance network, as proposed by Canada.

The final compromise agreement was shaped by the staunch opposition of industrialized countries to include provisions on mandatory assessed contributions, even as most developing countries argued that they would not accept a treaty that did not include binding provisions on assistance (Downie 2003). Article 12 thus commits developed country parties to provide "timely and appropriate technical assistance" to developing countries. It also stipulates that developed countries should provide "new and additional financial resources" to developing countries to help fulfill their different obligations under the Stockholm Convention. In addition, the GEF was designated the main financing mechanism on an interim basis until the first COP where industrialized countries and developing countries would review the efforts by the GEF and continue negotiations on funding and capacity building issues.

Implementing the Stockholm Convention

After the Stockholm Convention was adopted in 2001, parties and observers met in 2002 and 2003 to prepare for implementation.[11] The treaty entered into force in 2004. Since then, four COPs have been held, with the most recent one organized in 2009.[12] The EU and 163 countries were parties to the Stockholm Convention as of late 2009. Notable nonparties include the United States and Russia (which have also not ratified the CLRTAP POPs Protocol and the Rotterdam Convention). During the COPs, the parties have focused on three sets of issues: developing organizational structures and specifying country requirements, evaluating additional chemicals for possible regulations, and issues of technical and financial assistance. Many decisions on these issues, as well as continuing disagreements, are shaped by a multitude of governance and actor linkages across the chemicals regime.

Organizational Issues and Specifying Country Requirements
Many organizational issues reflect similar debates in other forums, particularly under the Rotterdam Convention. Sometimes these similarities facilitated decision making under the Stockholm Convention, as the Stockholm Convention parties could build on precedent-setting agreements they had already reached as Rotterdam Convention parties. However, linkages between treaties also meant that differences and unresolved policy issues were diffused from one policy forum to the other.

COP1 decided that the secretariat should be located within UNEP Chemicals in Geneva (Earth Negotiations Bulletin 2002c, 2005b). COP1 also finalized the composition of the POPs Review Committee. In contrast to the heated discussions on the membership of the Rotterdam Convention's Review Committee (see chapter 5), countries relatively easily settled on a membership structure for the POPs Review Committee by more or less copying the model agreed to under the Rotterdam Convention (Earth Negotiations Bulletin 2005b). There are thirty-one members on the review committee: eight from Africa, eight from Asia and the Pacific, three from Central and Eastern Europe, five from Latin America and the Caribbean, and seven from Western Europe and others. Membership rotates every two years. Meetings, which are held in between the COPs, are open to observers and experts at the invitation of the review committee.

Many developing countries again raised the issue of the working language of the review committee at COP1, as they had done at COP1 of the Rotterdam Convention less than a year earlier. For example, China argued for interpretation into all six UN languages to avoid a trade-off in participation of experts between those who held the most relevant expertise and those who were able to communicate best in English. In contrast, many industrialized countries once more pointed out the costs of translation services that would strain the convention budget. Based on a compromise that was slightly different from the one reached under the Rotterdam Convention, COP1 decided that all meetings of the POPs Review Committee would be translated into all UN languages as long as all meetings were held in Geneva, where the secretariat would hire translators at a minimum cost and avoid paying for travel costs (Kohler, 2006).

The Stockholm Convention calls for the use of regional centers to aid information sharing and transfer of funds and technologies to developing countries and countries with economies in transition (article 12). During the treaty negotiations, many industrialized and developing countries supported integrating these Stockholm Convention activities into the work programs of Basel Convention regional centers to capture synergies between capacity-building efforts across treaties. However, a few developing countries, including Syria and Brazil, advocated for the creation of separate centers for the Stockholm Convention (Earth Negotiations Bulletin 2002c, 2003, 2005b). In at least some cases, this was related to countries' desire to host such a Stockholm Convention regional center, as the Basel Convention regional centers were set up in other countries in their respective regions.

The establishment of the regional centers is based on self-nominations by countries within the five standard UN regions. There is no formal limit on the number of centers, but it is up to the countries within each region to assess their own needs. By 2009, twelve centers had been proposed across all five UN regions. At COP4, parties approved eight of these, while leaving decisions on other centers to subsequent COPs (Earth Negotiations Bulletin 2009).[13] Of the eight approved at COP4, two were already operating as Basel Convention regional centers (and two of the other four that have been proposed are also Basel Convention regional centers). This creates important organizational linkages between the Basel and Stockholm conventions on capacity-building issues. The use of separate and joint centers also raises questions about efforts to streamline management activities across treaties.

Parties are working to develop a global monitoring program to evaluate implementation progress. This involves improving POPs data sets, especially in developing regions (Earth Negotiations Bulletin 2006c). Governments are also developing and updating technical guidelines for the environmentally sound management of stockpiles and wastes (Earth Negotiations Bulletin 2009). The work on wastes is carried out in close collaboration with the Basel Convention secretariat, as the Basel Convention COPs are simultaneously formulating technical guidance documents on POPs wastes, linking activities under the two treaties (Earth Negotiations Bulletin 2005b, 2006c, 2007a, 2009). Parties are furthermore developing toolkits for the application of best available techniques and best environmental practices for controlling by-products, including dioxins and furans (Earth Negotiations Bulletin 2009). In addition, parties are engaged in discussion with the Rotterdam Convention secretariat on issues relating to the legal and illegal trade in POPs.

To facilitate the inclusion of chemicals under the treaty, countries during the treaty negotiations agreed that the Stockholm Convention should allow regulated exemptions on production and use. The development of the Register of Specific Exemptions and procedures for reporting was a major issue during the first few COPs (Earth Negotiations Bulletin 2003, 2005b). The register, which is maintained by the secretariat, is used to identify parties that have been granted an exemption to use a POP listed in annex A or B of the Stockholm Convention. The register lists each individual exemption given to each party on a substance-by-substance basis.

All exemptions listed in the register expire after five years, but the COPs may grant an extension for a maximum of another five years. When there are no longer any parties registered for a particular type of exemption, no new registrations can be made with respect to it.

Furthermore, sixteen countries by 2009 had issued production or use notifications under the separate DDT Register: Botswana, China, Ethiopia, India, Madagascar, Marshall Islands, Mauritius, Morocco, Mozambique, Myanmar, Senegal, South Africa, Swaziland, Uganda, Yemen, and Zambia. However, several additional countries also use DDT (Stockholm Convention Secretariat 2009a). Globally, approximately 5,000 metric tons (active ingredient) of DDT are used every year (Stockholm Convention Secretariat 2009b). Some of it is used as an active ingredient to produce dicofol, while most other uses are for malaria control. Three countries were known to still produce DDT by the early 2000s (Stockholm Convention Secretariat 2009a, 2009b). India was the largest producer, followed by China and South Korea. China, however, has indicated that it may cease to regularly produce and use DDT while retaining the right to use it in case of a domestic outbreak of malaria.

India and China also export DDT as both a technical product and a formulated product for vector control. Some of their exports of DDT as a technical product are used to formulate DDT in other countries, including Ethiopia and South Africa. South Africa furthermore exports some of its formulated DDT to other African countries. Reflecting the determination by several parties to keep using DDT against malaria-carrying mosquitoes, COP4 concluded that countries that still use DDT for disease vector control may need to continue doing so until better local, cost-effective alternatives become available. Issues relating to the continuing use of DDT and the development of alternatives (both chemical and nonchemical) will continue being part of discussion by the COPs. These discussions are linked to ongoing DDT debates under the CLRTAP POPs Protocol and within the WHO and the GEF.

In addition, parties are discussing the development of a compliance mechanism (Bankes 2003). So far, parties have been unable to agree on how to deal with compliance issues (Earth Negotiations Bulletin 2007a). In particular, China, India, and Iran have refused to accept anything but a self-triggering mechanism (Earth Negotiations Bulletin 2009). These discussions are similar to those under the Basel and the Rotterdam conventions, where opposition from leading industrialized and developing countries has prevented the establishment of strong compliance mechanisms (the Basel Convention mechanism created in 2002 is weak by design, and the Rotterdam Convention parties have yet to reach any kind of meaningful agreement on compliance issues). Future decisions on a compliance mechanism under one of these treaties will have significant implications for how compliance issues are addressed under the others as well.

Evaluating Additional Chemicals

The POPs Review Committee, which always gathers in Geneva under the agreement reached at COP1, has met once a year since its establishment at COP1.[14] Meetings are well attended, as the review committee's work is a critical part of implementing the convention and improve global protection against POPs. For example, in addition to the members of the review committee, representatives of thirty-nine countries and twenty-four NGOs attended the third meeting in 2007. Similarly, fifty-seven non-committee member countries and forty-three NGOs were present for the fourth meeting in 2008.

Based on the detailed five-step procedure for evaluating additional chemicals outlined in the Stockholm Convention, several countries were quick to propose new chemicals. The WWF furthermore in 2005 circulated a list of twenty additional chemicals that the organization deemed suitable for regulation under the Stockholm Convention (Earth Negotiations Bulletin 2005b). As of 2008, twelve chemicals had been proposed by parties for evaluation by the review committee. Not surprisingly, most of these proposals have come from European countries backed by the EU. This reflects their relatively stringent regional controls and is also consistent with their parallel initiatives on adding chemicals under the CLRTAP POPs Protocol. There is much interplay and overlap between the Stockholm Convention and the CLRTAP POPs Protocol with respect to regulated and proposed chemicals (see table 7.4).

In 2005, five chemicals were proposed for review: chlordecone and hexabromobiphenyl by the EU, lindane by Mexico, pentabromodiphenyl ether by Norway, and perfluorooctanesulfonate (PFOS) by Sweden. After the full assessment process, the review committee recommended the inclusion of all five in the Stockholm Convention (Earth Negotiations Bulletin 2007a).

Table 7.4
POPs included or proposed under the CLRTAP POPs Protocol and the Stockholm Convention by 2009

Chemicals	CLRTAP	Stockholm	Proposed CLRTAP	Proposed Stockholm
Aldrin	√	√		
Chlordane	√	√		
Chlordecone	√	√		
DDT	√	√		
Dicofol			√	
Dieldrin	√	√		
Dioxins	√	√		
Endosulfan			√	√
Endrin	√	√		
Furans	√	√		
Heptachlor	√	√		
Hexabromobiphenyl	√	√		
Hexabromocyclododecane			√	√
Hexachlorobenzene	√	√		
Hexachlorobutadiene			√	
Hexachlorocyclohexane	√	√√√		
Mirex	√	√		
Octabromodiphenyl ether		√	√	
PCB	√	√		
Pentabromodiphenyl ether		√	√	
Pentachlorobenzene		√	√	
Pentachlorophenol			√	
PFOS		√	√	
Polychlorinated naphthalene			√	
PAHs	√			
SCCP			√	√
Toxaphene	√	√		
Trifluraline			√	
TOTAL	16	21	12	3

Note: Under the CLRTAP POPs Protocol, hexachlorocyclohexane is listed as one set of substances and is stated to include lindane (gamma-hexachlorocyclohexane). Under the Stockholm Convention, the three hexachlorocyclohexane preparations (alpha-hexachlorocyclohexane, beta-hexachlorocyclohexane, and gamma-hexachlorocyclohexane/lindane) count as three separate chemicals.
Source: Updated based on H. Selin and N. Selin (2008).

In 2006, five more substances were nominated for review: alpha-hexa-chlorocyclohexane and beta-hexachlorocyclohexane by Mexico and octabromodiphenyl ether, pentachlorobenzene, and SCCP by the EU. All of these substances with the exception of SCCP were eventually recommended for regulation by the review committee (Earth Negotiations Bulletin 2007a, 2008c). The EU proposed endosulfan in 2007, one year before the EU proposed to regulate it under the CLRTAP POPs Protocol. In addition, Norway nominated hexabromocyclododecane in 2008, the same year Norway also proposed it under CLRTAP. Both these chemicals, together with SCCP, were still under review by the review committee by early 2009.

Consequently, the assessment work carried out by the review committee during its first four meetings between 2005 and 2008 resulted in the recommendation to control nine additional chemicals: pentabromodiphenyl ether, chlordecone, hexabromobiphenyl, lindane, PFOS, alpha-hexachlorocyclohexane, beta-hexachlorocyclohexane, pentachlorobenzene, and octabromodiphenyl ether. In a major decision, the parties at COP4 in 2009 unanimously agreed to add all of these chemicals to the Stockholm Convention. Eight of the chemicals were included in annex A, and one chemical was listed under annex B (see table 7.5). To make it possible to reach a consensus-based agreement on the inclusion of all nine substances, however, exemptions on the continued production or use had to be given for four of the chemicals considered by the parties at COP4: octabromodiphenyl ether (annex A), pentabromodiphenyl ether (annex A), lindane (annex A), and PFOS (annex B) (Earth Negotiations Bulletin 2009).

The flame retardants octabromodiphenyl ether and pentabromodiphenyl ether are no longer in production but exist in many plastic and foam rubber products. Country-specific exemptions allow recycling of products containing these chemicals. Country-specific exemptions on lindane requested by several developing countries allow its use as a human health drug against head lice and scabies. As both industrialized countries and developing countries argued that there were no cost-effective and environmentally friendly alternatives to many uses of PFOS, the parties agreed on a multitude of general exemptions. These include using PFOS in photo imaging, firefighting foam, insect baits, metal plating, leather and apparel, textiles and upholstery, paper and packaging, and rubber and plastics. At the same time, COP4 abolished several exemptions originally listed in the Register of Specific Exemptions. In addition, COP4 endorsed the establishment of a PCBs elimination network run by the Stockholm Convention secretariat in collaboration with the Basel Convention secretariat.

Although the addition of nine chemicals in 2009 was significant, there are signs of growing controversy as the review committee and COPs are increasingly tasked to consider chemicals still in widespread use. These chemicals have a greater economic value than the POPs that were originally included in the Stockholm Convention. For example, the cases of SCCP and endosulfan demonstrate that parties are not always able to reach an easy consensus. Major parties like China and Japan, together with the United States and the Chlorinated Paraffin Industry Association, argue that SCCP, which is still subject to a number of industrial uses, does not meet the environmental risk criteria in annex E (Earth Negotiations Bulletin 2007b). Regulations of the pesticide endosulfan, which is used extensively in agriculture, are similarly contested. At the review committee meeting in 2008, India and China, two leading producers and

Table 7.5
Chemicals listed in the Stockholm Convention by 2009

Annex A	Annex B	Annex C
Aldrin	DDT	Dioxins
Alpha-hexachlorocyclohexane	PFOS	Furans
Beta-hexachlorocyclohexane		Hexachlorobenzene
Chlordane		PCBs
Chlordecone		
Dieldrin		
Endrin		
Heptachlor		
Hexabromobiphenyl		
Hexachlorobenzene		
Lindane/gamma-hexachlorocyclohexane		
Octabromodiphenyl ether		
Mirex		
Pentabromodiphenyl ether		
Pentachlorobenzene		
Toxaphene		
PCBs		

users, led an effort to stop further evaluation (Earth Negotiations Bulletin 2008c).

There are multiple indications that the assessments by the POPs Review Committee—and the decisions by the COPs—are becoming more politicized (Earth Negotiations Bulletin 2006d, 2009). Many developing countries argue that resource inequalities between industrialized and developing countries hamper their ability to effectively participate in the review committee, negatively affecting the credibility, legitimacy, and salience of its work and recommendations. Discussions on assessments and expanding the number of regulated chemicals, including endosulfan and SCCP, are furthermore linked with parallel debates under the Rotterdam Convention (see chapter 5) and the CLRTAP POPs Protocol (see chapter 6). This is because many chemicals are simultaneously considered in multiple forums with overlapping membership. This gives raise to much linkage politics, involving both forum shopping and scale shopping by leader states seeking to expand regulations across policy forums.

Issues of Technical and Financial Assistance

As is the case with the other major chemicals treaties, issues relating to technical and financial assistance and capacity building are critical under the Stockholm Convention. Developing countries continuously stress that they are not able to meet their commitments—and effectively address domestic POPs problems—without additional support from donor countries and IGOs as well as the secretariat.

To aid countries in their development of national implementation plans, the IOMC organized a guiding workshop in 2002 (Earth Negotiations Bulletin 2002c). UNEP and the GEF furthermore held ten regional workshops between 2001 and 2003 (UNEP Chemicals, 2004). These workshops, funded by the GEF with contributions from Canada, Sweden, and Switzerland, were designed to assist developing countries in strengthening their national chemicals management programs toward the effective implementation of the Stockholm Convention. Several of these workshops were organized in collaboration with the Basel Convention regional centers. The parties also discussed formats for reporting domestic implementation activities to be periodically submitted to the secretariat. In addition, the GEF has financially supported the development of several national implementation plans (Earth Negotiations Bulletin 2002c).

Countries have furthermore developed a capacity assistance network. Several donor countries have given voluntary contributions to a "POPs Club" fund supporting activities under the Stockholm Convention (Earth Negotiations Bulletin 2002c). The GEF Council shortly after the Stockholm Convention was adopted made POPs one of its "focal areas," to support implementation. Under its fourth replenishment period, from 2006 to 2010, the GEF allocated US$300 million to POPs management (Earth Negotiations Bulletin 2007a). Many countries, however, fear that available resources are not enough to meet funding demands and call for increased contributions. At the same time, the stalemate persists between industrialized countries and developing countries on issues of mandatory versus voluntary financial contributions, dating back to the treaty negotiations (Earth Negotiations Bulletin 2009). Financial assistance and capacity-building issues are also connected to the establishment of a compliance mechanism and the operation of the Stockholm Convention regional centers.

Stockholm Convention funding debates are influenced by parallel discussions under the Basel and Rotterdam conventions (as well as in other environmental issue areas). Disagreements between industrialized and developing countries are basically the same in all chemicals forums. The fact that these debates are politically and practically linked means that a change in one forum (e.g., industrialized countries' acceptance of mandatory financial contributions under the Stockholm Convention) is hard to achieve in part because it would have a direct and significant influence on discussions (and possibly policy outcomes) also under other treaties. That is, industrialized countries are weary of setting a precedent in one policy forum because they realize that it would have ramifications far beyond that forum. At the same time, leading developing countries argue that increased funding is a prerequisite for accepting a compliance mechanism (Earth Negotiations Bulletin 2009).

Conclusion

The creation of the Stockholm Convention was the result of several scientific and political actions on POPs. The POPs issue did not emerge on the global agenda simply because the science was so compelling, or because of particular political interests, but because of a combination of scientific assessment work and political cooperation. In this respect, scientific and political POPs agendas are intimately co-constructed. Industrialized countries possessing relatively much human and scientific resources played dominating roles during the assessments and convention negotiations. However, developing countries and countries with economies in transition were also deeply involved. Furthermore, the constructive inclusion of less scientifically advanced countries, which also often lack economic and human resources for effective chemicals management, is critical for continuing legitimate policy making and effective implementation.

The regulation of nine new chemicals by the COP in 2009 was a significant step in implementing the Stockholm Convention. Nevertheless, the parties still face several political and management issues. Major challenges include expanding the number of regulated chemicals, establishing a comprehensive monitoring system, developing technical guidelines for managing wastes and reducing emission of by-products, generating funding for capacity building, operating the Stockholm Convention regional centers and determining their relationship with the Basel Convention regional centers, and creating a comprehensive compliance mechanism. Parties will keep working on these issues over many future COPs in parallel under the Basel, Rotterdam, and Stockholm conventions. Many of these issues are, of course, are closely interrelated.

Effective monitoring and the generation of assessment data are critical for measuring progress in addressing the POPs problem. At the same time, assistance to developing countries and countries with economies in transition is required to improve management capabilities. The Stockholm Convention regional centers linking global, regional, and domestic management efforts may facilitate capacity building. In addition to the issues that are specific to the effective operation of Stockholm Convention, countries and stakeholders face the challenge of ensuring uniform implementation of related life cycle provisions across treaties. Therefore, secretariats and parties must act to coordinate regulatory and administrative efforts under the Stockholm Convention with decisions and activities under the other major chemicals agreements to fulfill the goals of SAICM and ultimately strengthen multilevel governance.

8

Multilevel Governance and Chemicals Management: Past, Present, and Future

Improving the management of hazardous chemicals is an important sustainable development issue. Although some progress toward chemicals safety can be noted, many chemicals continue to pose unacceptable environmental and human health risks. This chapter analyzes multilevel governance issues as they relate to the creation and future of the chemicals regime. The chapter begins by summarizing the main components of the chemicals regime. It then returns to the three research questions posed in chapter 1, focusing on issues of coalitions, diffusion, and effectiveness. Next, the chapter discusses four multilevel governance challenges critical for increasing regime effectiveness and environmental and human health protection. The chapter ends with a few remarks regarding some of the major lessons that the chemicals case offers other governance efforts with respect to characters and implications of institutional linkages for strengthening multilevel governance.

The Chemicals Regime

The chemicals regime, created over several decades to mitigate harm to human health and the environment, focuses on global, regional, national, and local aspects of the chemicals problem. The treaties and programs that constitute the core of the regime address the transboundary transport of emissions and the international trade in hazardous chemicals and wastes, as well as domestic management challenges. These cross-scale issues are, of course, intimately linked: frequent long-range transport of emissions and extensive trade in chemicals and wastes add to significant local contamination and management problems across the world. The main chemicals treaties and programs assign states a shared responsibility to ensure that domestic activities do not cause damage to other states or areas outside national jurisdiction. This creates a legal foundation for addressing a wide range of chemicals issues and their associated management problems across governance scales.

While all countries face important management challenges, the chemicals regime recognizes the particular problems of developing countries and countries with economies in transition. The two POPs treaties also explicitly identify the vulnerability of Arctic ecosystems and indigenous communities from the contamination of toxic, persistent, and bioaccumulating chemicals. Several principles and norms are guiding policy making and management, including the principle of common but differentiated responsibilities. Based on this principle, for example, developing countries and countries with economies in transition are given more time to phase out the use of regulated chemicals and deal with contaminated equipment. Furthermore, treaties stipulate that industrialized countries have a responsibility to help developing countries and countries with economies in transition to develop their domestic management capacities through financial and technical assistance.

The chemicals treaties acknowledge the importance of precaution in guiding collective policy making and regulation, although the practical implications of applying the precautionary principle for risk assessment and management are frequently contested. The polluter-pays principle is included in the Stockholm Convention, giving legal recognition to the notion that the polluter should, in principle, bear the primary cost of mitigating pollution. For the parts of the regime that address the trade in commercial chemicals and wastes, the PIC principle has been established as a core legal principle identifying states' rights and responsibilities. Under the PIC schemes, a regulated chemical or waste cannot be exported from one country to another without the explicit consent of the importing state. At the same time, treaties stipulate that regulatory actions should not distort international trade and investment by discriminating against particular states or products.

The principles of common but differentiated responsibilities, precaution, polluter pays, and PIC are connected to norms on how to best manage hazardous chemicals. Since the 1960s, international cooperation has been expanded based on a belief shared by leading IGO, states, and NGOs that effectively managing the international trade in chemicals and wastes as well as reducing transboundary transport of emissions requires extensive political, technical, and scientific collaboration. Chemicals treaties also contain much normative language about the importance of the transfer of technology and other resources to developing countries and countries with economies in transition. As a result, there is a strong normative commitment to collectively deal with hazardous chemicals from global to local levels. Many treaty commitments to support regional and domestic capacity building are, however, formulated in rather weak language, putting few mandatory requirements on countries.

Based on the principles and norms guiding international cooperation and problem solving, the different treaties that constitute the core of the regime introduce life cycle regulations of a small set of hazardous chemicals, covering their production, use, trade, and disposal. The Stockholm Convention is the only treaty that focuses on all parts of the life cycle; the other treaties cover only part of it. Legal rules stipulate that parties should ban or restrict the production and use of POPs pesticides and industrial chemicals regulated by the two POPs treaties. Rules regarding the international trade of commercial POPs have also been established. Whereas the trade in some POPs is banned under the Stockholm Convention, the trade in other POPs pesticides and industrial chemicals is allowed. The trade in several hazardous chemicals is also subject to a detailed PIC procedure outlined in the Rotterdam Convention. Similarly, the Basel Convention stipulates PIC requirements for the trade in hazardous wastes.

The chemicals treaties contain mandatory provisions for the environmentally sound management, transport, and disposal of used chemicals and chemical wastes. Parties have a duty to reduce releases from stockpiles and wastes of chemicals controlled under the regime, based on a normative commitment that such releases should ultimately be eliminated. States are obligated to design secure domestic management practices for recycling and discarding used or unwanted chemicals. In addition to the regulatory focus on commercial chemicals, the POPs treaties outline requirements and recommendations for minimizing emissions of by-products. To this end, parties are required to apply the best available techniques and best environmental practices on major sources emitting these by-products. Stipulations for the source-specific application of best available techniques and best environmental practices are outlined in technical annexes to treaties.

The chemicals regime encompasses an open-ended approach to chemicals management, as regime participants believe that continuing assessments and regulations of additional chemicals beyond those currently covered by the different treaties are needed. To this end, treaties are designed so that regulations on commercial chemicals, by-products, and waste categories can be expanded without having to renegotiate entire treaties or having to develop new ones. Instead, detailed procedures for the work of chemical review committees linking scientific assessment with the political negotiation of expanded controls are included in the treaties (in the case of the CLRTAP POPs Protocol, a separate agreement was negotiated and adopted alongside the protocol). A few chemicals have already been added through these mechanisms, and many more are lined up for assessment and possible regulations.

Decision-making authority is invested in the COPs of the three global agreements, while the CLRTAP Executive Body performs this function for the CLRTAP POPs Protocol. This includes final decisions about the regulation of additional chemicals. All four treaties are administered by separate secretariats. Tasks assigned to these secretariats include arranging the meetings of the COPs and subsidiary bodies, providing administrative functions, and coordinating activities with other organizations (including other secretariats and science advisory bodies that are evaluating chemicals considered for regulation). Parties are required to develop implementation plans and periodically review and update these plans, as well as regularly submit implementation-related information to the secretariats. In this respect, the secretariats act as important clearinghouses for the exchange of scientific, technical, and policy data. Secretariats are also tasked with reviewing implementation progress and support national activities.

The chemicals treaties recognize the need for expanding international and domestic monitoring and collaborative research efforts to aid continuing scientific assessments and policy making (including expanding the number of regulated chemicals and wastes). The treaties state that such monitoring and research-oriented efforts should, among other things, focus on emission sources and environmental releases, emission transport patterns, contamination levels and trends in humans and the environment, and effects on human health and ecosystems. Treaties also note the importance of improving the availability of socioeconomic assessment data to inform policy developments and regulatory designs. In addition, leading IGOs, including UNEP, UNITAR, FAO, and the GEF in collaboration with the secretariats, are involved in a host of capacity-building activities, including through the regional centers established under the Basel and Stockholm conventions.

Coalitions, Diffusion, and Effectiveness

The history and complexity of the chemicals regime give rise to many interesting analytical issues and research topics. Chapter 1 posed three research questions focusing on the respective themes of coalitions, diffusion, and effectiveness. This section draws on empirical information and arguments from earlier chapters to address these questions.

How do coalitions of regime participants form in support of policy expansions, and how are their interests and actions affected by institutional linkages?
Coalitions of states, IGOs, and NGOs have played many influential roles in the creation and expansion of the chemicals regime. The same group of regime participants has not consistently been the driving force behind the many policy and management developments on hazardous chemicals since the 1960s; rather, different coalitions have been formed around specific sets of issues that coalition members have prioritized and promoted. In seeking particular policy expansions, coalitions of regime participants often engage in linkage politics as they connect discussions and policy making on related issues debated in multiple forums. Such linkages may have a discernable impact on participants' interests and activities as they seek to promote policy change through the strategic use of governance and actor linkages.

Coalitions of regime participants are formed around shared beliefs and interests, which are often shaped by actor and governance linkages. Coalitions are often led by states exhibiting intellectual and material leadership to promote specific issues or outcomes. The formation of coalitions is facilitated by the fact that representatives of many states use personal and professional connections to exchange ideas and knowledge on which common interests and policy positions may be formed. Because many of the issues addressed by the different chemicals treaties are so similar, the

same leading state representatives commonly meet under multiple treaties. This is reflected in the decision by states that the negotiations of the
Stockholm Convention were not to begin until the Rotterdam Convention was concluded, because government officials had the ability to negotiate only one major chemicals treaty at the time.

Attempting to build coalitions and generate political momentum for
policy change, leader states also often draw on the support and resources
of IGOs. Many IGOs and secretariats play important roles as coalition
partners and intellectual and material leaders by, for example, preparing
and hosting meetings, acting as nodes for information gathering and dissemination, and supporting capacity building. Furthermore, UNEP hosts
the two secretariats for the Basel and Stockholm conventions, while the
Rotterdam Secretariat is run jointly by UNEP and FAO. The CLRTAP
Secretariat located within the UNECE oversees the implementation of the
POPs Protocol. In addition, NGOs can be important coalition members
by advocating for particular policy positions, generating and diffusing
knowledge, and raising international and domestic awareness. These actions may help leader states and IGOs achieve their goals.

The fact that parallel and related discussions on many policy issues
going on in multiple forums—including the application of principles for
cooperation and regulation, the design of controls of specific chemicals,
and the development of mechanisms for monitoring, compliance, financial assistance, and capacity building—means that coalition members
must pay attention to all of these discussions, as well as consider how
one debate or decision may shape others. That is, a policy decision in one
forum can have significant influence on how a similar issue is addressed
in another forum. The occurrence of parallel debates in two or more forums also opens up multiple avenues for advocacy and policy change. In
fact, cross-venue coalitions are sometimes formed when a policy issue is
blocked or moving slowly in one venue and coalition members are looking to find alternative routes for forwarding their interests and generate
policy change.

More specifically, a coalition of UNEP, most African countries, the
Nordic countries, Greenpeace, and the Basel Action Network led early
policy developments on the trade in hazardous wastes and chemicals. This
coalition, believing that trade controls were needed particularly to help
developing countries reduce domestic problems, forwarded their cause

by linking debates and policy developments on wastes and chemicals. Yet coalition members failed to achieve a complete or partial trade ban in the face of strong opposition from a coalition of most industrialized countries and industry organizations, instead settling for voluntary PIC schemes. The anti-ban coalition also recognized the importance of institutional linkages; they realized that settling the trade issue on either chemicals or wastes would have important ramifications for the management of the other. This made them reluctant to compromise on either issue.

The pro-ban coalition continued to push for stricter trade controls, leading to the adoption of the Basel Convention. The Basel Convention, however, did not regulate the waste trade beyond making the PIC scheme mandatory because of continuing opposition from the anti-ban coalition. An expansion of the pro-ban coalition to include a greater number of developing countries and the EU, however, led to the adoption of the Ban Amendment at COP3, prohibiting hazardous wastes transports from industrialized to developing countries. While the EU has banned exports of hazardous wastes to developing countries, the Ban Amendment has yet to enter into force owing to resistance from industrialized countries such as Canada, Japan, and the United States (the United States has not ratified the Basel Convention either), industry organizations, and some developing countries that benefit from the financially lucrative waste trade.

Drawing on their experiences with policy achievements and failures under the Basel Convention, developing countries were heavily involved as coalition members in policy developments on chemicals during the 1990s. Continuing their close collaboration with UNEP and other IGOs, as well as environmental NGOs such as the Pesticide Action Network, Greenpeace, and WWF, developing countries were leading proponents of turning the voluntary PIC mechanism on chemicals trade into a legally binding instrument. They wanted to do this to strengthen its position under international law and increase legal demands on exporters. To this end, they once again linked policy developments on hazardous wastes and chemicals by using policy developments under the Basel Convention as leverage for going from a voluntary to a mandatory PIC scheme also on the chemicals trade. Developing countries remain very active in the implementation of the Rotterdam Convention, including efforts to expand the PIC list.

Developing countries were furthermore active coalition members in the process leading to the creation of the Stockholm Convention. On POPs, however, most early intellectual and material leadership came from northern industrialized countries. Many of the issues relating to coalition politics visible during the development and implementation of the Basel Convention and the Rotterdam Convention can also been seen with respect to the creation of the CLRTAP POPs Protocol and the Stockholm Convention. Similar to how the coalition behind expanding trade regulations on wastes and chemicals strategically used policy developments in one forum as leverage to advocate for the adoption of similar policies in another forum, northern industrialized countries used CLRTAP as a way to both expand regional POPs regulations and scale up the POPs issue pushing for a global POPs treaty.

On POPs, particularly Canada and the EU (including individual member states) have been critical in building coalitions with other states, IGOs, and NGOs. These actions are driven by a shared interest to reduce transboundary transport of emissions. Hazardous chemicals are an important human health and environmental issue in Canada and the EU, and their actions reflect their relatively stringent domestic and regional controls. In Canada, the POPs issue is closely linked with scientific and political concerns about Arctic environmental contamination and human health risks, especially for indigenous populations. Indigenous peoples' rights became a hot political issue in Canada in the 1980s, and the inclusion of indigenous groups in scientific and political work increased domestic sensitivity to chemicals issues, leading Canada to be a strong advocate of international regulations on POPs to address long-range transport of emissions.

The EU shared many Canadian concerns and was also the main initiator of the 2020 goal on safe chemicals production and use adopted at the WSSD in 2002. In fact, the WSSD goal was based on an almost identical objective in the EU's sustainable development strategy, formulated in 2001. This strategy identifies chemicals management as a major sustainable development issue and declares the intention of all EU member states to, "by 2020, ensure that chemicals are only produced and used in ways that do not pose significant threats to human health and the environment" (European Commission 2002, 35). The EU's Sixth Environment Action Programme, which operates from 2002 to 2012 and provides the environmental component of the sustainable development strategy, sets additional priorities, including expanding risk assessments and management strategies, and substituting hazardous chemicals with safer chemicals or alternative technologies. The EU is pushing similar issues under SAICM.

Canada and the EU have exercised combinations of intellectual and structural leadership on international chemicals management in several ways, including advocating for regional and global policy expansions in combination with a readiness to host and finance international political and scientific meetings, initiating and sponsoring environmental assessments, and funding national experts in international organizations preparing background documents and policy proposals. The willingness of Canada and Sweden to initiate and financially support scientific assessments on POPs in CLRTAP in the early 1990s and under the IFCS in the late 1990s was instrumental for drawing attention to issues and consequences of long-range transport of emissions. This was critical for building a coalition pushing for the creation of the CLRTAP POPs Protocol and then using this work to initiate the development of the Stockholm Convention.

As important coalition members, groups of Arctic indigenous peoples also strategically used the many governance and actor linkages between the CLRTAP POPs Protocol and the Stockholm Convention to advance their cause. In these efforts, they were supported by Arctic countries and many IGOs and environmental NGOs. The CLRTAP POPs Protocol acknowledges in the preamble that "the Arctic ecosystems and especially its indigenous people, who subsist on Arctic fish and mammals, are particularly at risk because of the biomagnification of persistent organic pollutants." The ICC later referred to this statement to gain similar recognition under the Stockholm Convention, which in the preamble states that "the Arctic ecosystems and indigenous communities are particularly at risk because of the biomagnification of persistent organic pollutants and that contamination of their traditional foods is a public health issue."

Finally, the chemicals regime draws attention to several coalition issues that can be important also in other issue areas. Many examples from the chemicals regime demonstrate that participants may use repeated interactions to discover and formulate common interests and policy positions. These shared interests and positions can form the basis for coalition building around particular policy issues. Relationships between different coalitions can be contentious, as different groups express diverging views and support opposing policy outcomes. Coalition building and actions to promote specific policy developments are shaped by actor and governance linkages, which influence regime participants' interests and give rise to linkage politics (including forum and scale shopping). Institutional linkages also raise political stakes in that the effects of a policy decision may have important consequences for decision making in other forums as well.

How do regime participants diffuse regime components across policy venues, and how are policy diffusion and expansion efforts shaped by institutional linkages?

Under the chemicals regime, many identical or similar principles, norms, and rules can be found embedded in multiple treaties with overlapping participation of a wide range of states, IGOs, and NGOs. The adoption of many of these principles, norms, and rules has been driven by coalitions of regime participants championing specific issues as the scope and stringency of the regime have been gradually expanded, demonstrating intellectual and material leadership. Many of these policy developments have also been the result of coalitions of regime participants purposefully diffusing ideas, knowledge, and policy proposals across policy forums, with a noticeable impact on decision making and the implementation of commitments and management programs across the regime.

Policy diffusion efforts by coalitions of regime participants are shaped by institutional linkages in several ways. Sometimes policy diffusion is aided by the fact that participants share an interest in harmonizing principles, norms, and rules across treaties so that they are not faced with conflicting commitments, which would impede treaty implementation and weaken regime effectiveness. The desire to coordinate activities was a major driver behind the development of SAICM and can also be seen with respect to the actions by parties to the CLRTAP POPs Protocol and the Stockholm Convention to ensure that chemicals covered by both treaties are subject to compatible regulations. Similarly, parties to the Basel Convention are mindful of policy developments on wastes under the Stockholm Convention as they develop technical guidelines for the disposal of POPs wastes. Governments also work on harmonizing national reporting requirements and formats under all major chemicals treaties.

In general, a growth in governance and actor linkages opens up new avenues for coalitions of states, IGOs, and NGOs to promote policy change. Some policy diffusion initiatives may be driven by interest-based calculations by some regime participants to advance specific policy changes that may not be equally desirable to all stakeholders. This strategic use of policy diffusion frequently involves some kind of forum or scale shopping (or both). That is, regime participants strategically consider which policy forum they think is most likely to respond in a favorable way before presenting new ideas, knowledge, and policy proposals or act to expand an issue from one governance scale to another. In an institutionally dense issue area where there is much interdependence between policy instruments, the intent is often not only to achieve policy change in one forum but to use these changes to shape policy outcomes in multiple policy forums.

There are many instances of forum and scale shopping behavior under the chemicals regime. The selection of a forum may be based on combinations of interests and influence. For example, developing countries and other coalition members wanting to transform the voluntary PIC procedure on chemicals and wastes into legally binding commitments elected to work through UNEP to generate momentum for policy change because they believed UNEP would be receptive to their demands. Because of the regional composition of the fifty-eight members elected to serve four-year terms on the UNEP Governing Council, developing countries have been able to exercise a great deal of influence on the formulation of UNEP declarations and decisions. On chemicals and wastes issues, developing countries, in collaboration with supportive industrialized countries, used UNEP in the 1970s and the 1980s to promote the institutionalization of a PIC procedure and the creation of the Basel and Rotterdam conventions.

Similarly, Canada, Sweden, and the other countries championing the POPs issue selected CLRTAP to pursue the development of a POPs treaty because they believed that the UNECE would be supportive of this effort. They picked CLRTAP in part because no global IGO expressed any interest in taking up the POPs issue in the late 1980s, but also because many of the same countries promoting the POPs issue in the past had successfully used CLRTAP to create protocols on sulfur and nitrogen emissions and hoped that they could carry over their influence on these issues to shape the development of POPs controls. These countries also used CLRTAP as a stepping-stone for promoting the creation of the Stockholm Convention with the aid of UNEP, the IFCS, and the WHO. Forum and scale shopping continues as parties consider which additional chemicals to propose for regulatory action under which treaty based on where they think they are most likely to get their desired outcome.

Both cognitive interaction (in which coalitions transfer ideas and knowledge from one policy forum to another) and interaction through commitment (in which parties' acceptance of specific principles, norms, and rules in one policy forum affect stakeholders' interests and decision making in another forum) are important under the chemicals regime. Co-alitions of regime participants frequently cognitively link different instruments and use the acceptance of particular ideas in one treaty as a means to shape debates and argue for their insertion in other ones. For example, coalitions led by developing countries have pushed for the diffusion of the PIC principle and the principle of common but differentiated responsibilities across chemicals treaties. Similarly, the EU, with the support of many developing countries, has been a vocal supporter of the inclusion of the precautionary principle in treaties and declarations, trying to shape decisions and policy outcomes in multiple forums.

Another prominent example of cognitive interaction can be found in the extensive transfer of POPs assessment reports and data across IGOs as well as from the CLRTAP POPs process to the global negotiations. This transfer was pushed by CLRTAP parties and other supporters of more stringent international controls on POPs. Such actions by leader states and organizations serve to perceptually link regional and global actions on POPs that have been in development since the 1990s. As a result, the framing of the POPs issue as one of long-range transport posing a particular threat to the Arctic region carried over from the regional POPs negotiations, shaping the global framing of the POPs issue. Similarly, much of the work on establishing screening criteria for assessing additional POPs diffused from CLRTAP to the global negotiations, resulting in very similar assessment criteria under the two POPs treaties even if they were negotiated independently and at different times.

The diffusion of regime components shaping decision making also frequently takes place via interaction through commitment. For example, the coalition supporting stronger trade regulations on both chemicals and wastes used the acceptance of the PIC procedures for the trade in hazardous wastes as leverage to establish a similar scheme on commercial chemicals. Commitments agreed on by countries under the CLRTAP POPs Protocol shaped their interests and positions during the global POPs negotiations as they worked to establish compatible principles, rules, and assessment mechanisms. There are also many similarities between the regulation of specific POPs under the CLRTAP POPs Protocol and the Stockholm Convention. Ongoing assessments of additional chemicals across treaties are closely linked by regime participants seeking to shape future POPs policy as well as the listing of additional chemicals on the Rotterdam Convention PIC list.

Cases of cognitive interaction and interaction through commitments under the chemicals regime involve much science-policy interplay. Successful scientific assessments and transfer of assessment data across forums are dependent not only on the scientific credibility of the assessment reports but also on their policy salience and political legitimacy. It is often harder to ensure the acceptance and appropriate use of scientific assessments in diverse stakeholder groups and multiple forums as opinions of what is credible, salient, and legitimate information may differ greatly. However, the POPs case demonstrates that it is possible. POPs assessment data generated under CLRTAP played an important part in the selection of the original twelve chemicals covered by the Stockholm Convention, as well as the preparatory work preceding treaty negotiations. Issues of credibility, salience, and legitimacy are also critical to the successful operation of the chemical review committees and continued decision making by the COPs.

Nevertheless, the existence of governance and actor linkages does not guarantee smooth policy expansions. Efforts by a coalition to diffuse specific principles and assessment data from one forum to another can cause frictions, as seen in debates surrounding the precautionary principle. Difficulties with policy expansions may also be compounded by differences in membership. For example, the transfer of POPs assessments and regulations from CLRTAP to the global process was complicated by the fact that most countries involved in the global efforts had not been involved in the regional work. As a result, they did not share the previous experience with those countries that were involved in the development of the CLRTAP POPs Protocol. CLRTAP parties, aware of this, therefore supported regional workshops before the negotiations on the Stockholm

Convention, in part to explain earlier work on POPs to representatives from countries dealing with the POPs issue for the first time.

Political disagreements in one forum sometimes spill over into others and may hinder policy diffusion and decision making. This was, for example, visible in the lengthy debates between opposing coalitions around the mandate and operation of chemical review committees under the Rotterdam and Stockholm conventions. On these issues, linkages between the parallel discussions raised the political stakes. All regime participants realized that a decision under one of the treaties would have direct implications for how the issue was settled under the other treaty. This led to protracted conflicts across policy venues. Furthermore, disagreements in the chemical review committee and COP of one treaty, as discussions and decisions are characterized by much science-policy interplay, also affect the work of review committees and COPs associated with other treaties. This can be seen in the case of endosulfan and other chemicals addressed under both POPs agreements and the Rotterdam Convention.

Other controversies diffused from one policy forum to another because parties treat them as interlinked can be found in ongoing debates regarding the establishment of more comprehensive mechanisms on monitoring, compliance, and capacity building. These discussions are carried out separately under each agreement as their legal bases are found in the specific language of each of the four treaties. Parties are, for example, not trying to create one monitoring mechanism operating throughout the regime, but are discussing how to deal with monitoring problems within the context of each treaty. Yet these debates are linked as parties recognize that an agreement under one treaty will shape debates and outcomes elsewhere. Again, this makes regime participants hesitant to compromise in one forum as they know that would have ramifications for how their interests may or may not be met under other treaties.

Finally, the kinds of diffusion issues that have become increasingly common under the chemicals regime can be expected to become more and more important in other issue areas as well. That is, it is very likely that the diffusion of ideas, knowledge, and policy will become commonplace in a host of regimes where there is growing institutional density. Much of this diffusion will be driven by coalitions of regime participants that seek to take advantage of actor and governance linkages to promote their interests. To this end, coalition leaders frequently engage in linkage politics to take advantage of the overlap in membership and participation in different policy forums, which creates a multitude of channels for the diffusion of principles, norms, and rules across forums. As a result, practitioners and analysts across issue areas need to pay closer attention to how linkage politics may both facilitate and hinder governance efforts.

How do institutional linkages influence the effectiveness and design of multilevel governance efforts?
Global governance of hazardous chemicals is shaped by a multitude of institutional linkages influencing the operation of the chemicals regime across global, regional, national, and local governance scales. That is, the many horizontal and vertical governance and actor linkages between major treaties and management programs affect the implementation of each of these treaties and programs, as well as the effectiveness of the regime to introduce comprehensive life cycle regulations to mitigate environmental and human health problems stemming from hazardous chemicals. Many (but not all) of these linkages appear to be synergetic rather than conflictual. This is in large part because regime participants recognize the importance of governance and actor linkages and have taken a series of political and administrative measures to enhance legal and regulatory consistency across treaties.

Many impact-level interactions (where the governance target of one policy forum is affected by spillover effects from decisions and actions taken in another forum), as well as some behavioral interaction (where behavioral changes by stakeholders in one policy forum affect the effectiveness of another forum), can be seen in the chemicals regime. On issues of impact-level interaction, the chemicals treaties share a multitude of governance overlaps, as policy developments and management activities under one treaty frequently affect the ability of other treaties to fulfill their stated objectives. For example, most chemicals controlled under the regime are regulated under two or more treaties. Furthermore, many chemicals assessed by different chemical review committees are considered in tandem under multiple treaties. Therefore, debates and regulatory decisions under one treaty will continue to influence the operation of other treaties, as well as the effectiveness of the regime.

A multitude of management linkages are critical to impact-level interaction and regime effectiveness. Each chemicals treaty has its own organizational structure. The mandates of subsidiary bodies operating under each treaty come explicitly from the text of that treaty, and the subsidiary bodies also report exclusively to their respective COPs. Yet the expansion of management activities with respect to issues such as legal and administrative support, information sharing, the development of technical guidelines, and running capacity-building programs affects outcomes outside the forum where specific decisions and actions are taken. The growth in policy and management linkages creates a continuing need for regime participants to carefully assess how related policy developments and management efforts bear on implementation across forums. It also highlights the importance of regime participants' continuing to focus on harmonizing policy and management activities.

Cases of behavioral interaction are not dominant under the chemicals regime but draw attention to the fact that the changing behavior of regime participants in one policy forum may push participants in another forum to adapt their behavior. For example, by allowing prominence to Arctic indigenous groups during the protocol negotiations, the CLRTAP parties set an important precedent that also shaped the way these groups were able to influence the global assessments and negotiations of the Stockholm Convention. Furthermore, changing attitudes toward the use of DDT against malaria-carrying mosquitoes by the WHO and among parties to the CLRTAP POPs Protocol and the Stockholm Convention shape public debates and policy positions on DDT use and controls in all these forums. Similarly, a changing behavior toward precaution or the utilization of scientific assessments in one policy forum may affect debates and attitudes under multiple treaties.

The successful operation of the chemical review committees under the Stockholm Convention, the CLRTAP POPs Protocol, and the Rotterdam Convention are critical for long-term regime effectiveness. This is also an area where there is much behavioral interaction and impact-level interaction, shaping the production of usable knowledge. Perceptions of what constitutes scientifically credible, policy salient, and politically legitimate information in these review committees and how committee recommendations are judged by the COPs have a direct impact on the regulation of additional chemicals. Because of governance and actor linkages, controversies and decisions in one forum influence policy outcomes under other treaties as well. This is especially true when the same chemical is assessed simultaneously in two or more treaties with an overlapping set of parties. Therefore, the effectiveness of one treaty mechanism for adding chemicals is very much dependent on the effectiveness of the others.

As regime participants have come to acknowledge the importance of institutional linkages, they have paid more attention to the design of governance structures and efforts that better link treaties and programs seeking to boost regime effectiveness. These design-related efforts attempt both to avoid overlapping and counterproductive regulations and management actions and capture synergies. This desire to benefit from regulatory and management synergies was a major driving force behind the development of SAICM, as well as the creation of the Ad Hoc Joint Working Group to Enhance Cooperation and Coordination among the Basel, Rotterdam, and Stockholm Conventions. This working group, which held its first meeting in 2007, had forty-five members (fifteen from each COP). The working group, at its third meeting in 2008, issued a set of recommendation to be considered by the respective COPs to the Basel Convention, the Rotterdam Convention, and the Stockholm Convention.

Regime participants considering these and other proposals for designing better governance structures are faced with a spectrum of options, ranging from merging separate treaties and subsidiary bodies into a single legal and organizational structure, to establishing a series of smaller mechanisms for coordinating activities of separate treaties and bodies. Most political and practical actions within the chemicals regime on design issues have been toward the latter end of the spectrum. That is, regime participants are primarily looking at ways to enhance the workings of existing treaties and their associated organizational structures rather than reshaping the entire regime. Nevertheless, this requires examining a host of important design issues and may involve changing the ways in which important decision-making and management activities are structured and carried out (see table 8.1.).

One major design issue concerns the organization of meetings by the COPs and subsidiary bodies. Historically, meeting schedules have been set within the context of each treaty. This resulted in spreading out meetings over calendar years and geographical regions with little consideration for how meetings under separate treaties were practically linked. Based on a recommendation by the Ad Hoc Joint Working Group to Enhance Cooperation and Coordination among the Basel, Rotterdam, and Stockholm

Table 8.1
Major design areas and issues for enhancing regime effectiveness

Design Area	Design Issue
• Organization of meetings	• Coordinate meetings of COPs and legal, political, and scientific subsidiary bodies across treaties.
• Management support	• Coordinate legal and administrative support by treaty secretariats and international organizations across forums.
• Reporting	• Coordinate and streamline procedures and requirements for national reporting across treaties and programs.
• Information sharing	• Establish procedures for generating and sharing scientific, technical, and policy information across forums.
• Capacity building	• Coordinate efforts for strengthening management capacity across forums.
• Resource mobilization	• Generate funds for strengthening management capacity and coordinating funding activities across forums.
• Monitoring and compliance	• Establish mechanisms for monitoring and compliance across treaties.

Note: Developed from Stockholm Convention Secretariat (2006), Basel Convention Secretariat (2006), and Ad Hoc Joint Working Group on Enhancing Cooperation and Coordination among the Basel, Rotterdam, and Stockholm conventions (2008).

Conventions, a common COP for the three treaties to address mainly administrative issues will be held in 2010. However, holding back-to-back meetings of COPs and legal, political, and scientific subsidiary bodies, including chemical review committees, may give rise to new concerns; for example, more intense meeting periods increase demands on country representatives before and during such periods, which many smaller countries with limited human and financial resources may not be able to meet.

Regime participants also focus on ways in which management support provided by the secretariats, which share many of the same core functions, can be better coordinated. Traditionally parties have had to seek legal and administrative advice on domestic implementation from different secretariats, although many of the same issues and chemicals are addressed by multiple treaties. This has resulted in unnecessary duplication of work by secretariats and makes it complicated for parties to get advice on overlapping policy issues. Similarly, many countries are calling for more harmonization of national reporting procedures and requirements. Countries are struggling to meet all reporting demands because reporting formats on specific issues and chemicals differ among treaties. More streamlined support services provided by secretariats and national reporting across forums clearly make sense.

There are compelling reasons to establish more coordinated and comprehensive procedures for gathering scientific, technical, and policy information to aid management across forums. This issue is closely related to the design of the chemical review committees. The fact that the review committees operate on regulated schedules with respect to the holding of the COPs facilitates predictability and continuity in their work. Taking geographical representation into consideration improves their integrity and enhances the credibility, salience, and legitimacy of their work. So far the parties have taken few practical steps to ensure that there is disciplinary diversity among committee members who generally work in national governments. Studies suggest, however, that interdisciplinary representation with members coming from both governmental and nongovernmental organizations can enhance the effectiveness of science advisory bodies (Haas 2004, Kohler 2006).

Many multilevel governance challenges stem from problems of cognitive and practical gaps between the global and the local. It is often hard to ensure that global policy and management efforts fit a wide range of local circumstances and needs. At the same time, local people often have little voice in the development of global policy, which is ultimately intended to change their behavior. One of the more interesting developments under the chemicals regime that may address some cognitive and practical gaps is the creation of the regional centers under the Basel and Stockholm conventions. These regional centers offer interesting and novel opportunities for organizationally linking policy-making and management activities within and across multiple governance scales. At the same time, the establishment of a growing number of regional centers under multiple treaties creates new coordination challenges.

An important task facing international and national policy makers and administrators collaborating through the Basel Convention and Stockholm Convention regional centers and other bodies is to assign specific management tasks across different forums and levels of social organization. Successful multilevel allocation of management tasks is not simply a question of identifying a single correct scale of governance. Rather, "in most cases, the key to success lies in allocating specific tasks to the appropriate level of social organization and then taking steps to ensure that cross-scale interactions produce complementary rather than conflicting actions" (Young 2002, 266). In other words, horizontal and vertical linkages influence not only the allocation of particular management tasks but also the operation of management programs. The effectiveness of management programs is also dependent on the reviewing and updating of activities under these programs.

The operation of effective management and capacity-building programs on hazardous chemicals is affected by a multitude of horizontal and vertical linkages across the regime. Horizontally there are compelling reasons to ensure that political and technical efforts on chemicals and waste management organized by secretariats and IGOs are supportive and do not overlap. This both saves financial and human resources and helps parties meet related obligations under more than one treaty. Vertically many management tasks across treaties could be coordinated through existing and future regional centers. These include sharing scientific, technical, and socioeconomic information; training local people in the safe handling of hazardous chemicals and wastes; transferring cleaner technology; expanding public-private partnerships; supporting national data collection and reporting; and improving public awareness and education.

Multilevel governance requires resource mobilization: having an architect design an elaborate building but not giving enough money to the builders will not produce a structurally sound home. Developing countries and secretariats routinely point out that many management and assistance programs are underfunded, with perpetual discussions at COPs about raising additional resources. Some improvements generating and allocating financial resources for capacity building may be gained by designing new means for raising and pooling resources across treaties and GEF funds rather than having parallel discussions—and disagreements—at each COP. Some of these resources could be channeled through the Basel Convention and Stockholm Convention regional centers. Nevertheless, redesigning management efforts is not enough; they must be both politically and financially supported to work.

Many issues relating to the successful design of more comprehensive treaty mechanisms for monitoring and compliance also depend heavily on the political will of the parties, coupled with the willingness of countries and aid agencies to raise resources for their implementation. Monitoring and compliance issues are central to regime effectiveness, and such mechanisms could cover multiple treaties. This would help ensure regulatory consistency across the regime, as well as reduce the burden on parties to compile and submit multiple reports. Both industrialized and developing countries, however, are reluctant to give up sovereignty and approve the design of independent monitoring and compliance mechanisms. At the same time, many developing countries are likely to struggle to find human and financial resources to meet expanded data-gathering and reporting requirements.

Finally, many effectiveness and design issues important under the chemicals regime are consequential in other regimes too. In issue areas where there are multiple treaties and other kinds of policy-making forums, decisions and behavioral changes in one forum will have an impact on the effectiveness of other policy and management efforts. As regimes experience growing institutional density, participants must consider how policy expansions across forums affect regime effectiveness. This includes paying close attention to both horizontal and vertical linkages in the design of successful structures for cross-scale management. To this end, debates and experiences under the chemicals regime may be helpful, including on issues of coordination of meetings and activities by secretariats, improving information sharing and reporting, enhancing capacity building and resource mobilization, and expanding monitoring and compliance efforts.

Four Multilevel Governance Challenges

Although the chemicals regime is much more developed and encompasses a greater number of regulations and management programs now than in the 1980s, many critical governance challenges remain. We are still far from having achieved global chemicals safety. Realizing this, the CSD has made chemicals and waste management one of its key themes in 2010–2011 to promote much-needed progress. States, IGOs, and NGOs also continue to work on a host of regulatory and management issues under each chemicals treaty as well as under SAICM. Building on the earlier discussion, this section examines four multilevel governance issues critical to increased regime effectiveness: (1) enhance ratification and implementation of existing regulations, (2) expand risk assessments and controls, (3) improve management capacity and raise awareness, and (4) minimize generation of hazardous chemicals and waste (H. Selin 2009b).

Enhance Ratification and Implementation of Existing Regulations
Enhanced ratification of global and regional treaties would improve the status of the chemicals regime and also be beneficial because of the many horizontal and vertical linkages between agreements and management programs. Of course, having a large number of ratifications and parties does not necessarily mean that the chemicals regime is effective (that related treaties and programs help reduce environmental and human health problems caused by hazardous chemicals) or that levels of compliance are high (that parties adhere to provisions and regulations set out in treaties and programs). Even when there are many parties, countries may give the treaty little priority or lack domestic resources to implement and enforce commitments. Nevertheless, a lack of ratification sends a message that a country does not believe that an agreement is important or that it disagrees with its objectives and requirements. In contrast, a high degree of participation indicates that governments take an issue seriously.

Although the vast majority of the world's countries have ratified the Basel Convention, the United States (one of the world's largest producer and trader of hazardous wastes) and some medium-sized countries, including South Korea, are not parties. In addition, the Basel Convention Ban Amendment has not received sufficient ratification to enter into force, and even fewer countries have signed the Basel Protocol on Liability and Compensation. More than 160 countries are parties to the Stockholm Convention, and almost 130 countries have ratified the Rotterdam Convention. The United States and several developing countries, however, are not parties to these two treaties. Twenty-nine countries have ratified the CLRTAP POPs Protocol, but the United States has yet to become a party. Increased ratification of all four treaties would increase the number of states that take on formal responsibilities as well as strengthen the treaties' position under international law.

The fact that treaties cover partially different life cycle issues raises important issues about regulatory and management synergies. The benefits of closer cooperation between the different COPs and secretariats in many areas of regime development are clear. This could not only help save limited resources, but also make it easier to improve regulatory and management consistency across agreements and minimizing (if not eliminating) the risk that important issues fall in regulatory or administrative cracks between treaties (Downie, Kruger, and Selin 2005). This, however, is made difficult by the uneven membership across the major chemicals treaties—another reason to promote increased ratification (there will, of course, always be differences in membership between regional and global treaties). Better coordination of policy making and implementation-related activities across treaties would also make it easier for parties to meet their commitments.

The stringency of mandates and levels of resources given to secretariats are important for treaty effectiveness. Studies show that secretariats can be particularly important when parties have weak domestic administrative capacities (Wettestad 2001). The secretariats, in collaboration with regional centers, could operate more ambitious multilevel mechanisms for monitoring and enforcement. Countries' statements and actions during COPs and other meetings, however, demonstrate that many remain protective of national sovereignty. Most industrial and developing countries are unwilling to give much independent authority to treaty bodies on issues of data collection and monitoring, and they are even less willing to give secretariats or other treaty bodies the right to take action against a party that does not fulfill its obligations. Nevertheless, there is reason to believe that stronger secretariats with access to more resources would aid implementation and compliance with agreements across the regime.

In promoting implementation, it is important to have a comprehensive life cycle focus. Ensuring that current bans and restrictions on production and use of commercial chemicals are enforced is critical to reduce environmental and human exposure (even if environmental contamination problems and health risks will linger for a long time due to the widespread environmental dispersion and persistence of many hazardous chemicals). Imposing trade bans and restrictions under the POPs treaties and ensuring the success of the PIC procedures under the Rotterdam and Basel conventions are necessary to limit problems with hazardous chemicals and wastes, particularly in many developing countries. The entry into force and implementation of the Basel Convention Ban Amendment would further-more strengthen the controls on waste transports. In addition, countries need to increase efforts to prevent illegal shipping of chemicals and dumping of e-wastes and other kinds of hazardous wastes.

Expand Risk Assessments and Controls

Expanding risk assessments and regulations beyond the small number of chemicals that are currently subject to life cycle controls is necessary. For most commercial chemicals, there are only scant data on emissions, environmental dispersion, and ecosystem and human health effects. Data gathering and assessments are costly, and there are compelling financial reasons to increase the scope of collective scientific and political efforts. This could also prevent duplication of assessments across forums and governance scales. Similarly, evaluations of candidate chemicals for possible controls under chemicals treaties are often long and arduous processes. There is a need to design mechanisms for conducting quicker risk assessments of a large number of chemicals to produce more policy-relevant information. To this end, the creation of an intergovernmental panel on chemical pollution, loosely modeled after the Intergovernmental Panel on Climate Change, has been proposed (Scheringer 2007).

The idea behind creating an intergovernmental panel of chemical pollution is that expanding assessments of hazardous chemicals on a needed scale requires international collaboration. Such a panel could build on existing global and regional initiatives to collect and synthesize scientific and socioeconomic data. Assessment reports published by the panel could help identify information gaps and targeting assessments on specific chemicals and regions. A more centralized assessment process could also facilitate consistent policy making across global and regional treaties. Before such a panel were to be created, however, its political and practical benefits should be carefully considered in relationship to anticipated costs of operation. Just because the Intergovernmental Panel on Climate Change has produced well-regarded reports and contributed to a better understanding about climate change, similar panels may not be appropriate in every environmental issue area.

Regulations on hazardous chemicals need to be expanded using the chemical review committees. Many POPs-like chemicals attract much scientific attention (AMAP 2009). For example, concentrations of polybrominated diphenyl ethers, extensively used as flame retardants, increased exponentially in Arctic seals between 1981 and 2000 (Ikonomou, Rayne, and Addison 2002). PFOS, which was included in the Stockholm Convention in 2009 but with numerous exemptions, accumulates in Arctic seals and polar bears (Giesy and Kannan 2001). Several countries continue to use the pesticide endosulfan. Although these chemicals are subject to increasing controls, many recent discussions in chemicals review committees and COPs are contentious because of the economic value of some chemicals proposed for regulation. While regime participants are working to expand regulations on POPs, assessments and policy making should be accelerated to evaluate the number of possible POPs that warrant attention.

Furthermore, the use of many non-POP pesticides is continuing largely unabated in many parts of the world. Even if these substances do not strictly meet the treaty criteria of a POP, they fall under WHO category I (extremely and highly hazardous) and category II (moderately hazardous) of dangerous pesticides (Mancini et al. 2005). Consequently, there is a need to consider additional pesticides for inclusion in the Rotterdam Convention PIC procedure, as well as increase participation by FAO, WHO, and UNITAR on pesticide management issues. This includes expanding efforts on integrated pest management, which involves using a combination of environmentally friendly methods designed to significantly reduce and, where possible, eliminate the use of pesticides. Such methods may require increased use of pest-resistant crop varieties, the employment of local insects for pest control, the design of more effective crop rotation schemes, and improved soil management.

The growing generation and importance of e-wastes highlights many connections between treaties, as there is a need to step up regional and global actions on e-waste. As the consumption of electrical and electronic goods is increasing in both industrialized and developing countries, levels of e-wastes are growing too. While the EU produces approximately 8.7 million tonnes of e-waste annually, China generated "only" roughly 1.7 million tonnes of e-waste in 2006 (Greenpeace 2008, Eugster et al. 2008). Chinese e-waste levels are, however, expected to increase to 5.4 million tones by 2015. Substantial amounts of e-wastes are also illegally imported into China for dismantling in small and unlicensed companies and workshops. These cases and the difficult situations in many other countries point to the need to expand domestic and international controls and enforcement efforts to better track and control the recycling and disposal of e-waste (Sonak, Sonak, and Giriyan 2008; Yang 2008).

The management of hazardous substances and waste is set to become even more closely intertwined with the creation of a global mercury treaty. After many years of assessments and debate, the UNEP Governing Council in 2009 decided to launch treaty negotiations with a focus on concluding an agreement in 2013. This treaty will build on long-standing regional action on mercury, including the CLRTAP Heavy Metals Protocol (N. Selin and H. Selin 2006). This protocol, which was adopted alongside the CLRTAP POPs Protocol in 1998 and entered into force in 2003, controls mercury, cadmium, and lead. Some countries have also expressed support for a global treaty that covers other heavy metals as well. These policy efforts on heavy metals are linked with the chemicals regime in many ways, as a large number of e-waste and other waste categories contain hazardous chemicals and heavy metals. Many regional treaties also regulate both hazardous chemicals and heavy metals.

Improve Management Capacity and Raise Awareness
To mitigate environmental and human health problems from hazardous chemicals throughout their life cycle, it is necessary to improve regional and domestic management capacities and raise awareness all over the world. Although no country has perfect management structures, many developing countries in particular have difficulties ensuring safe use of hazardous pesticides and industrial chemicals, as well as the environmentally sound management of hazardous wastes. Furthermore, the continuing release of by-products such as dioxins, furans, and PAHs poses significant management challenges. Many countries lack access to the best available techniques and best environmental practices for limiting emissions of these and other by-products from industrial manufacturing, combustion processes, and waste management as outlined in technical annexes and guidelines developed under the two POPs treaties.

To improve management capacities and raise awareness requires better linking of global, regional, national, and local forums and activities. International assistance is required in many cases to improve domestic conditions. To this end, the Basel Convention and Stockholm Convention regional centers can help coordinate capacity-building and training activities across treaties and governance scales. The regional centers can play many important roles, including information generation and sharing, border controls, and building of domestic management capabilities for emission prevention and remediation of contaminated sites. On these issues, the regional centers are set up to function as critical organizational nodes linking capacity-building efforts by IGOs such as UNEP, FAO, WHO, UNITAR, the secretariats, and activities and the needs of local authorities and handlers of chemicals and wastes.

Of course, effective operation of the regional centers and successful operation of capacity building and training programs require human, financial, and technical resources. Yet generating adequate resources to run such programs efficiently continues to be a problem. So far only small, inadequate resources have been committed to the regional centers by IGOs, Northern donor countries, and member countries in each region. Contentious budget discussions under both the Basel and Stockholm conventions do not indicate that many more resources will be made available soon. This is unfortunate, as it is clear that the life cycle management of hazardous chemicals would benefit from the ability of the regional centers to act more forcefully on a range of management and enforcement issues.

National implementation plans submitted by parties under the Stockholm Convention, together with numerous scientific and socioeconomic studies, show that better chemicals management also includes raising awareness among farmers and industrial workers through education about chemical hazards. Such education programs should be coupled with better training of users in selecting a safe chemical control strategy based on which specific chemical is used. Key issues for the safe handling of chemicals include the secure handling of industrial chemicals and the appropriate application of pesticides for targeting pests and vector-borne diseases, the practical use of hand-operated and power-operated equipment, and the wearing of protective gear (WHO 2006). Yet awareness raising and training are not always enough. The authors of a recent study of acute pesticide poisoning among cotton growers in India concluded:

The extent of pesticide poisoning among farmers and workers in developing countries is worrying. In the extreme hot weather of the tropics, protective gear does not seem to be a viable solution to eliminate occupational risks. Educating farmers about the pesticide hazard alone has not achieved significant results. The solution seems to be in the replacement of pesticides with non- or less toxic alternatives. One example of such alternatives can be found in the integrated pest management approach. (Mancini et al. 2005, 231–232)

The India study illustrates that even when farmers are educated about health risks and protective gear is available, safety equipment may still not be used. This is because available safety equipment can be prohibitively expensive for rural farmers and may be cumbersome to use in hot weather in tropical and subtropical regions. Because of this situation, IGOs, states, and NGOs must not only promote the use of protective clothing and make it more readily available for farmers in poor areas, but they must also support the development of new gear more appropriate for tropical conditions. The authors of the India study, together with many other experts, also argue that efforts on promoting chemicals safety should be carried out in the context of expanding the use of integrated pesticide management practices in regions and countries that are still lagging in their application (Mancini et al. 2005; Nwilene, Nwanze, and Youdeowei 2008).

Minimize Generation of Hazardous Chemicals and Waste

Ultimately the most effective way to protect human health and the environment from hazardous chemicals is to avoid using them in the first place. Regulatory efforts have historically focused on the management of known or suspected hazardous chemicals rather than promoting the development of less harmful chemicals or nonchemical alternatives. Furthermore, in traditional procedures for managing hazardous chemicals, the burden of proof is on regulators to prove that a chemical is not safe rather than the producer or seller having to produce data demonstrating that a substance is not likely to cause adverse environmental and human health effects. Assessments and policy developments under the chemicals treaties and SAICM continue to proceed down these traditional tracks, which are slow and reactive. Only through the development and application of quicker and more proactive procedures for assessment and regulation can the main chemicals treaties become truly effective.

One country at the forefront of progressive chemicals policy is Sweden (Löfstedt 2003). As early as 1969, Sweden adopted a law introducing a reversed burden of proof by requiring firms to demonstrate the safety of activities that may be environmentally dangerous. The substitution principle—requiring firms to collaborate with public authorities to assess the possibility of replacing a hazardous chemical with a less hazardous alternative—became part of Swedish chemicals regulation in 1973. Currently, "a nontoxic environment" is one of sixteen national environmental quality objectives. Since joining the EU, Sweden has worked with other member states, the European Commission, and the European Parliament to design a more efficient and precautionary system for assessing and chemicals. A main outcome of this process is the Regulation on Registration, Evaluation, Authorisation, and Restriction of Chemicals (REACH), which entered into force in 2007 (H. Selin 2007, 2009a).

While much national chemicals policy outside Europe remains reactionary and places the burden of proof on regulators, commercial handling of a chemical covered by REACH is prohibited unless it is demonstrated to be harmless, adequately controlled, or that societal benefits outweigh costs. REACH furthermore explicitly targets all substances that are CMR (carcinogenetic, mutagenic, and toxic for reproduction), PBT (persistent, bioaccumulative, and toxic), or vPvB (very persistent and very bioaccumulative). All chemicals that exhibit these inherent characteristics—rather than on the basis of proven harm—have to be individually authorized by European authorities before they can be sold on the EU market. REACH provides a framework for a more precaution-based approach to chemicals management. It also contains guidelines for the phaseout and substitution of hazardous chemicals to less harmful substances or nonchemical alternatives.[1]

There is furthermore a need to focus more on waste minimization to protect the environment and reduce human health problems in all parts of the world. Parties to the Basel Convention have largely developed mechanisms to control the transboundary movement of wastes and technical guidelines for waste management, paying little attention to treaty stipulations on waste minimization. Despite a series of political declarations since the 1970s that the generation of hazardous wastes needs to be reduced, levels of hazardous wastes, including discarded chemicals or wastes containing chemicals, continue to increase in industrialized and developing countries. A reduction in waste generation could be reached in part through the application of cleaner production methods as well as the adoption of more proactive requirements for waste prevention.

Rapidly growing levels of e-waste—and the development of national, regional, and global efforts to handle it—further connect issues of hazardous chemicals and waste management (H. Selin and VanDeveer 2006, Yang 2008). The introduction of governmental regulations and market-based incentives that expand producer responsibility and make firms responsible for their products—including electronic and electrical goods—throughout their entire life cycle could play a significant role in stimulating more effective waste minimization efforts. E-waste is also attracting more attention under the Basel Convention. For example, the Mobile Phone Partnership Initiative was established as a public-private partnership to promote the environmentally sound management of old mobile telephones. This and other similar efforts, however, remain limited in scope and are entirely voluntary.

EU policy developments on producer responsibility and e-waste are again ahead of global standards on e-waste management. The Directive on the Restriction of the Use of Certain Hazardous Substances in Electrical and Electronic Equipment (RoHS) and the Directive on Waste Electrical and Electronic Equipment (WEEE) are designed to phase out several hazardous substances from electronic goods and increase the recycling of such goods, reducing the amount of e-waste going to final disposal (H. Selin and VanDeveer 2006). European consumers are required to return a large number of electronic goods to the producers, which are responsible for recycling, reprocessing, and safely disposing of the equipment and its components. In this respect, WEEE and RoHS, together with REACH, were created to increase incentives for producers to take waste management requirements into account when designing and producing products.

The influence of REACH, RoHS, and WEEE is seen outside Europe through processes of international economic integration and policy diffusion. Whereas many early international standards for consumer and environmental protection were set in the United States because of the relative stringency of early U.S. regulations in combination with the size of the U.S. economy, the EU is increasingly replacing the United States as the main setter of global product and assessment standards on hazardous chemicals. Interrelated EU policy developments of hazardous chemicals and wastes are attracting much attention from policy advocates and decision makers around the globe. Large producers and users of chemicals and electronic goods, such as China, Japan, South Korea, and California, are copying EU policy ideas in this area helping to raise standards and expand producer responsibility (H. Selin and VanDeveer 2006, Yang 2008, Renckens 2008, H. Selin 2009a).

In addition to having many environmental and human health benefits, a more proactive policy approach to chemicals and waste management reduces economic costs of cleaning up polluted areas. Although there are no reliable global data on such costs, the U.S. situation is an indication of the extent of the problem. The U.S. Comprehensive Environmental Response, Compensation, and Liability Act (commonly known as the Superfund Act) was passed in 1980 following the discovery of sites like the ones in Love Canal in upstate New York. In 2007, this act listed 275 priority chemicals and heavy metals commonly found at contaminated sites all over the country. The federal government spent $1.2 billion on Superfund projects in 2006 alone. Although twenty-four sites were completed in 2007, more than 1,250 sites on the Superfund National Priority List remain unaddressed, and more sites will be added in the future (H. Selin 2009b).

Finally, green chemistry—the use of principles that reduce or eliminate the use or generation of hazardous substances in the design, manufacture, and application of chemicals—is an effort to incorporate environmental and health concerns into the development of new chemicals (Anastas and Warner 1998). To target the chemicals problem at its source, green chemistry proponents stress the importance of synthesizing substances with little or no environmental toxicity. Chemicals should also be designed so that at the end of their functional lives, they break down into innocuous degradation products as part of a broader effort to create a more sustainable use of materials (Geiser 2001). Expanded public and private sector acceptance of green chemistry is a critical step toward ensuring the long-term safe production and use of chemicals. To this end, both voluntary industry-led programs like Responsible Care and mandates set by governments can play important roles.

Conclusion

As cooperative efforts to improve chemicals safety continue, the chemicals regime offers lessons for other policy areas with respect to institutional linkages and multilevel governance issues. With growth in vertical and horizontal linkages within issue areas, institutional density affects the interests and strategies of regime participants, including how they engage in linkage politics and form coalitions in support of or opposition to particular policy proposals. Regime participants operating in situations of a high degree of institutional linkages will think not only about how to advance their interests with respect to a specific policy issue within one forum, but also to consider how choices they make in that forum will influence their interests and policy outcomes in other policy arenas. This will have direct implications for regime creation and implementation as actions and decisions across forums become politically and practically linked.

As scholars and policy makers adjust to governance conditions where governance and actor linkages are playing a greater role and try to assess implications of growing institutional density, it should be recognized that institutional linkages can have both positive and negative effects on collective problem-solving efforts. In some cases, governance and actor linkages across policy forums can facilitate policy making, which may allow regime participants to capture important regulatory and management synergies. However, regime participants may also act in ways that hinder policy developments in response to governance and actor linkages between policy forums. In such cases, controversy between coalitions in one policy forum spills over into another forum, causing stalemate in both. Such an impasse may be more difficult to break because policy issues are linked across multiple policy forums, raising the overall political stakes.

An increase in institutional linkages increases the political and organizational need to better link management efforts across global, regional, national, and local governance scales. Only through integrated multilevel governance can many environmental issues be effectively addressed. On hazardous chemicals, the development of sets of goals and rules targeting the long-range transport of emissions and international trade, as well as the operation of a multitude of capacity-building programs, make chemicals management a complex cross-scale issue. Many other environmental issue areas share similar multilevel traits. As a result, environmental policy efforts need to combine the development of collective principles, norms, and rules with the design of localized management programs that allow flexibility in the use of specific implementation measures and instruments depending on local circumstances.

One seemingly promising option for achieving better governance from global to local scales is to expand the use of regionally based organizations for management and capacity building, as in the case of the Basel Convention and Stockholm Convention regional centers. These regional centers engage in many activities that are critical also outside the area of chemicals management, including human training and public education, technology transfer, and data collection and reporting. Numerous studies demonstrate that successful governance depends on clearly defined goals and standards that are periodically reviewed, and that implementation progress is assessed against these goals and standards. Furthermore, effective global governance must support regional and local capacity building based on many different conditions and needs, which are typically better identified locally and regionally than globally.

In these respects, participants in other issue areas are well advised to look at expanding multilevel management and capacity-building efforts under the chemicals regime, including through the regional centers. The use of these centers may also be a way to expand monitoring and compliance mechanisms and develop public-private partnerships. However, the chemicals case also clearly illustrates that while design improvements are important to capture regulatory synergies, they are not enough by themselves. Many regional centers are struggling to find the means to operate effectively. If regional centers are established in other issue areas but without the necessary resources, they will fall short of their potential. Improved multilevel governance of hazardous chemicals will be achieved only when adequate political and financial support is provided, and this holds true in any environmental policy area.

Notes

Chapter 1

1. The Blacksmith Institute and Green Cross Switzerland publish a Top Ten list and database of some of the world's most dangerous pollution problems and contaminated areas. For more information, see http://www.worstpolluted.org.

2. For more discussion about the regime concept and analysis, see chapter 2.

3. The ozone case is not included in this analysis of the chemicals regime. Although specific chemicals are responsible for the depletion of the stratospheric ozone layer, this issue is different from problems associated with hazardous chemicals at the earth's surface. In fact, there are stronger political and physical linkages between the ozone depletion and climate change issues than between stratospheric ozone depletion and the treaties covered in this book. However, treaties and organizations focusing on stratospheric ozone depletion are included in SAICM, which means there may be closer connections between policymaking on stratospheric ozone depletion, and other chemicals issues in the future.

Chapter 3

1. http://heritage.dupont.com/touchpoints/tp_1939/depth.shtml.

2. The name Agent Orange came from the colored identification band that was painted on the storage barrels. Other herbicides used during the Vietnam War were Agent Pink, Agent Green, Agent Purple, Agent White, and Agent Blue.

3. In fact, dioxin is not a single substance but a collection of congeners that share a similar structure. The most toxic and high profile of these is the congener 2,3,7,8-tetrachlorodibenzo-p-dioxin, also know as 2,3,7,8-TCDD.

4. The full name of the Superfund Act is the Comprehensive Environmental Response, Compensation, and Liability Act.

5. http://europa.eu.int/comm/environment/seveso/.

6. http://www.bhopal.com/chrono.htm.

7. The report was first presented at the American Chemical Society's 164th National Meeting in August and September 1972 and was published in 1975.

8. The following thirteen actions plan have been created under the Regional Seas Program: (1) Mediterranean Action Plan (adopted in 1975); (2) Red Sea and Gulf of Aden Action Plan (adopted in 1976, revised in 1982); (3) Kuwait Action Plan Action Plan (adopted in 1978); (4) West and Central African Action Plan (adopted in 1981); (5) Caribbean Action Plan (adopted in 1981); (6) East Asian Seas Action Plan (adopted in 1981); (7) South-East Pacific Action Plan (adopted in 1981); (8) South Pacific Action Plan (adopted in 1982); (9) Eastern Africa Action Plan (adopted in 1985); (10) Black Sea Strategic Action Plan (adopted in 1993): (11) North-West Pacific Action Plan (adopted in 1994); (12) South Asian Seas Action Plan (adopted in 1995); and (13) North-East Pacific Action Plan (adopted in 2001).

9. The fifteen countries were Armenia, Burkina Faso, Chile, Congo, Costa Rica, Côte d'Ivoire, Djibouti, Haiti, Georgia, Madagascar, Mongolia, Rwanda, São Tomé and Principe, Serbia, and Syria.

Chapter 4

1. "The Electronic Wasteland," 60 Minutes, November 10, 2008.

2. Six negotiations meetings were held in October 1987 (Budapest), February 1988 (Geneva), June 1988 (Caracas), November 1988 (Geneva), January and February 1989 (Luxembourg), and March 1989 (Basel).

3. The nine COPs were held at Piriapolis, Uruguay, December 1992; Geneva, March 1994; Geneva, September 1995; Kuching, Malaysia, February 1998; Basel, December 1999; Geneva, December 2002; Geneva, October 2004; Nairobi, November and December 2006; Bali, June 2008.

4. See http://www.ban.org/main/hall_of_shame.htm, accessed July 2008.

5. The fourteen regional centers are located in Argentina, China, Egypt, El Salvador, Indonesia, Iran, Nigeria, Senegal, Slovak Republic, Russian Federation, Samoa, South Africa, Trinidad and Tobago, and Uruguay.

Chapter 5

1. The negotiation meetings leading to the Rotterdam Convention were held in Brussels, March 1996; Nairobi, September 1996; Geneva, May 1997; Rome, October 1997; and Brussels, March 1998.

2. The seven FAO regions are thus different from the five broader UN regions. The seven standard FAO regions are Africa, Asia, Europe, Latin America and the Caribbean, Near East, North America, and Southwest Pacific.

3. The Montreal Protocol also bans both imports and exports of ozone-depleting substances to nonparties.

4. These five meetings were held in Rome, July 1999; Geneva, October and November 2000; Rome, October 2001; Bonn, September and October 2002; Geneva, November 2003; and Geneva, September 2004.

5. Four COPs had been held by 2008: Geneva, September 2004; Rome, September 2005; Geneva, October 2006; and Rome, October 2008.

6. The interim chemical review committee met five times: Geneva, February 2000; Rome, March 2001; Geneva, February 2002; Rome, March 2003; and Geneva, February 2004.

7. The chemical review committee had met five times by 2009: Geneva, February 2005; Geneva, February 2006; Rome, March 2007; Geneva, March 2008; and Rome, March 2009.

Chapter 6

1. The four meetings were held in Stockholm, March 1991; Port Stanton, Canada, May 1992; Berlin, May 1993; and The Hague, February 1994.

2. Forums that were deemed to have too narrow a geographical domain and membership included the Oslo-Paris Commission, the North Sea Task Force, the Helsinki Commission, and the Great Lakes abatement and control agreements. In contrast, the OECD had a much broader geographical coverage and membership, but its mandate was believed to be too restricted.

3. The four meetings were held in Geneva, March 1995; Geneva, July 1995; Geneva, May 1996; and Aylmer, Canada, October 1996.

4. All five meetings were held in Geneva and took place January 1997; June 1997; October 1997; December 1997; and February 1998.

5. PCP is, however, identified in the article on research, development, and monitoring as a substance that warranted special attention.

6. The nine chemicals regulated by both the Rotterdam Convention and the CLRTAP POPs Protocol are aldrin, chlordane, DDT, dieldrin, heptachlor, HCB, HCH/lindane, toxaphene, and PCBs. Under the Rotterdam Convention, HCH and lindane are counted as two separate substances, but not under the CLRTAP POPs Protocol. There are thus ten overlapping chemicals if HCH and lindane are counted separately.

7. The four meetings of the ad hoc Expert Group were held in The Hague, November 2000; Torun, Poland, October 2001; Geneva, June 2002; and Oslo, March 2003.

8. The six meetings were held in The Hague, May 2004; Prague, May and June 2004; Vienna; May and June 2005; Dessay, Germany, February 2006; Tallinn, May and June 2006; Vienna, June 2007.

Chapter 7

1. The ICC uses *Inuit* to refer to the Inupiat, Yupik (Alaska), Inuit, Inuvialuit (Canada), Kalaallit (Greenland), and Yupik (Russia).

2. The eight Arctic countries are Canada, Denmark (because of Greenland), Finland, Iceland, Norway, Russian Federation, Sweden, and the United States.

3. The five negotiation sessions were held in Montreal, June and July 1998; Nairobi, January 1999; Geneva, September 1999; Bonn, March 2000; and Johannesburg, December 2000. The two meetings of the Criteria Expert Groups were organized in Bangkok, October 1998, and Vienna, June 1999. The financial assistance meeting was held before the diplomatic conference in Stockholm, May 2001.

4. The seven POPs that were covered by both the Rotterdam and Stockholm conventions were aldrin, chlordane, DDT, dieldrin, heptachlor, hexachlorobenzene, toxaphene, and PCBs (endrin and mirex are not yet covered by the Rotterdam Convention).

5. In addition, parties are allowed several general exemptions that allow them to continue using existing items that contain small quantities of POPs and to produce and use very small quantities of POPs for laboratory and other closed purposes.

6. http://www.who.int/mediacentre/factsheets/fs094/en/index.html.

7. Annex D contains basic scientific criteria for persistence, bioaccumulation, potential for long-range transport, and adverse effects that a substance has to meet in order to qualify as a POP.

8. Each member of the review committee is appointed for four years that may be extended for a second term. The thirty-one members of the committee are appointed based on the principle of regional representation: eight African states, eight Asian and Pacific states, three central and eastern European states, five Latin American and Caribbean states, and seven Western European and other states.

9. Annex E outlines information needed to compile a risk profile, including production, use, and release data; environmental behavior and toxicological data; monitoring data; exposure data; national and international risk evaluations; and the status of the chemical under international treaties.

10. Annex F lists information needs, such as the efficiency of possible control measures, alternative products and processes, positive and negative societal impacts of regulations, waste and disposal implications, information availability and public education, and existing national or regional regulations.

11. The two meetings were held in Geneva in June 2002 and July 2003.

12. The COPs were held in Punta del Este in May 2005, Geneva in May 2006, Dakar in May 2007, and Geneva in May 2009.

13. The eight Stockholm Convention regional centers approved by COP4 are located in Brazil, China (also a Basel Convention regional center), Czech Republic, Kuwait, Mexico, Panama, Spain, and Uruguay (also a Basel Convention regional center). The four nominated Stockholm Convention regional centers not approved by COP4 are located in Algeria, Iran (also a Basel Convention regional center), Russia, and Senegal (also a Basel Convention regional center).

14. The four meetings to date were held in November 2005, November, 2006, November 2007, and October 2008.

Chapter 8

1. See also the REACH SIN (Substitute It Now) list compiled by the International Chemical Secretariat in collaboration with several other NGOs. This list is intended to push public agencies and firms to accelerate substitution of hazardous chemicals. http://www.chemsec.org/list.

References

Ad Hoc Joint Working Group on Enhancing Cooperation and Coordination among Basel, Rotterdam and Stockholm Conventions. 2008. *Report of the Ad Hoc Joint Working Group on Enhancing Cooperation and Coordination among Basel, Rotterdam and Stockholm Conventions on the Work of its Second Session.* UNEP/FAO/CHW/RC/POPS/JWG.2/18.

Africa Research Bulletin: Economic, Financial and Technical Series. 2006. *Cote D'Ivoire Dumping Ground* 43 (9):17107–17108.

Alston, Philip. 1978. International Regulation of Toxic Chemicals. *Ecology Law Quarterly* 7 (2): 397–456.

Anastas, Paul T., and John C. Warner. 1998. *Green Chemistry: Theory and Practice.* Oxford: Oxford University Press.

Arctic Monitoring and Assessment Programme. 2002. *Arctic Pollution 2002.* Oslo: AMAP.

Arctic Monitoring and Assessment Programme. 2003. *AMAP Assessment 2002: Human Health in the Arctic.* Oslo: AMAP.

Arctic Monitoring and Assessment Programme. 2004. *AMAP Assessment 2002: Persistent Organic Pollutants in the Arctic.* Oslo: AMAP.

Arctic Monitoring and Assessment Programme. 2009. *Arctic Pollution 2009.* Oslo: AMAP.

Asante-Duah, Kofi D., and Imre V. Nagy. 1998. *International Trade in Hazardous Waste.* New York: E & FN Spon.

Bäckstrand, Karin. 2001. *What Can Nature Withstand? Science, Politics and Discourses in Transboundary Air Pollution Diplomacy.* Lund: Lund Political Studies, 116.

Bankes, Nigel. 2003. The Stockholm Convention in the Context of International Environmental Law. In *Northern Lights against POPs: Combatting Toxic Threats in the Arctic,* ed. David Leonard Downie and Terry Fenge, 160–191. Montreal: McGill–Queen's University Press.

Basel Convention Secretariat. 2007. *Draft Report on the Operation of the Basel Convention Regional and Coordinating Centres.* December 14. Geneva: Basel Convention Secretariat.

Basel Convention Secretariat. 2006. *Recommendations on Improving Coopera-tion and Synergies Prepared by the Secretariat of the Basel Convention.* Geneva: Basel Convention Secretariat.

Béland, Pierre, Sylvain DeGuise, Christiane Girard, André Lagacé, Daniel Mar-tipeau, Robert Michaud, and Derek C. G. Muir 1993. Toxic Compounds and Health and Reproductive Effects in St. Lawrence Beluga Whales. *Journal of Great Lakes Research* 19 (4):766–775.

Betsill, Michele M., and Harriet Bulkeley. 2006. Cities and the Multilevel Gover-nance of Global Climate Change. *Global Governance* 12 (2):141–159.

Betsill, Michele M., and Elisabeth Corell, eds. 2008. *NGO Diplomacy: The Influ-ence of Nongovernmental Organizations in International Environmental Nego-tiations.* Cambridge, Mass.: MIT Press.

Boardman, Robert. 1986. *Pesticides in World Agriculture: The Politics of Interna-tional Regulation.* London: MacMillan.

Börzel, Tanja A. 2002. Pace-Setting, Foot-Dragging and Fence-Sitting: Member State Responses to Europeanization. *Journal of Common Market Studies* 40 (2):193–214.

Bowes, Gerald W., and Charles J. Jonkel. 1975. Presence and Distribution of Poly-chlorinated Biphenyls (PCB) in Arctic and Subarctic Marine Food Chains. *Journal of the Fisheries Research Board of Canada* 32 (11):2111–2123.

Breitmeier, Helmut, Oran R. Young, and Michael Zürn. 2006. *Analyzing Inter-national Environmental Regimes: From Case Studies to Database.* Cambridge, Mass.: MIT Press.

Breman, Joel, Martin S. Alilio, and Anne Mills. 2004. Conquering the Intoler-able Burden of Malaria: What's New, What's Needed: A Summary. [Supplement] *American Journal of Tropical Medicine and Hygiene* 71 (2):1–15.

Brickman, Ronald, Sheila Jasanoff, and Thomas Ilgen. 1985. *Controlling Chemi-cals: The Politics of Regulation in Europe and the United States.* Ithaca, N.Y.: Cornell University Press.

Brikell, Berndt H. 2000. Negotiating the International Waste Trade: A Discourse Analysis. PhD diss., Örebro University.

Brooks, Paul. 1972. *The House of Life: Rachel Carson at Work.* Boston: Hough-ton Mifflin.

Brown Weiss, Edith. 1993. International Environmental Issues and the Emergence of a New World Order. *Georgetown Law Journal* 81 (3):675–710.

Bryant, Lisa. 2007. Ivory Coast Still Suffering from Toxic Spill. *Voice of America*, December 15.

Caldwell, Lynton Keith. 1996. *International Environmental Policy: From the Twentieth to the Twenty-First Century.* 3rd ed. Durham, N.C.: Duke University Press.

Cambodia. 2006. *National Implementation Plan for the Stockholm Convention on Persistent Organic Pollutants.*

Carson, Rachel. 1962. *Silent Spring.* Cambridge, Mass.: Riverside Press.

Cash, David W., and Susanne C. Moser. 2000. Linking Global and Local Scales: Designing Dynamic Assessment and Management Processes. *Global Environmental Change* 10 (2):109–120.

Chambers, W. Bradnee. 2008. *Interlinkages and the Effectiveness of Multilateral Environmental Agreements*. Tokyo: United Nations University Press.

Chossudovsky, Evgeny. 1989. *East-West Diplomacy for Environment in the United Nations*. New York: UNITAR.

Clapp, Jennifer. 2001. *Toxic Exports: The Transfer of Hazardous Wastes from Rich to Poor Countries*. Ithaca, N.Y.: Cornell University Press.

Colborn, T., D. Dumanoski, and J. P. Myers. 1996. *Our Stolen Future*. New York: Dutton.

Coulter, Jeff. 1982. Remarks on the Conceptualisation of Social Structure. *Philosophy of the Social Sciences* 12 (1):33–46.

Datamonitor. 2005. *Chemicals in the United States: Industry Profile*. October.

Datamonitor. 2006. *Chemicals in Europe: Industry Profile*. December.

DeSombre, Elizabeth R. 2006. *Global Environmental Institutions*. New York: Routledge.

Dewailly, Eric, Albert Nantel, Jean-P. Weber, and François Meyer. 1989. High Levels of PCBs in Breast Milk of Inuit Women from Arctic Quebec. *Bulletin of Environmental Contamination and Toxicology* 43 (5):641–646.

Dewailly, E., A. Nantel, S. Bruneau, C. Laliberte, L. Ferron, and S. Gingras. 1992. Breast Milk Contamination by PCDDs, PCDFs and PCBs in Arctic Quebec: A Preliminary Assessment. *Chemosphere* 25 (7–10):1245–1249.

Dewailly, Eric, Pierre Ayotte, Suzanne Bruneau, Claire Laliberte, Derek C. G. Muir, and Ross J. Norstrom. 1993. Inuit Exposure to Organochlorines Through the Aquatic Food Chain in Arctic Quebec. *Environmental Health Perspectives* 101 (7):618–620.

Dewailly, Eric, and Christopher Furgal. 2003. POPs, the Environment, and Public Health. In *Northern Lights against POPs: Combatting Toxic Threats in the Arctic*, ed. D. L. Downie and T. Fenge, 3–21. Montreal: McGill–Queens University Press.

Donald, J. Wylie. 1992. The Bamako Convention as a Solution to the Problem of Hazardous Waste Exports to Less Developed Countries. *Columbia Journal of Environmental Law* 17 (2):419–458.

Downie, David Leonard. 2003. Global POPs Policy: The 2001 Stockholm Convention on Persistent Organic Pollutants. In *Northern Lights against POPs: Combatting Toxic Threats in the Arctic*, ed. David Leonard Downie and Terry Fenge, 133–159. Montreal: McGill–Queens University Press.

Downie, David Leonard, Jonathan Krueger, and Henrik Selin. 2005. Global Policy for Hazardous Chemicals. In *The Global Environment: Institutions, Law and Policy*, ed. Regina S. Axelrod, David Leonard Downie, and Norman J. Vig, 125–145. Washington, D.C.: CQ Press.

Dreher, Kelly, and Simone Pulver. 2008. Environment as "High Politics"? Explaining Divergence in US and EU Hazardous Wastes Export Politics. *Review of European Community and International Environmental Law* 17 (3):308–320.

Earth Negotiations Bulletin. 1997a. *Report of the Third Session of the INC for an International Legally Binding Instrument for the Application of the Prior Informed Consent Procedure for Certain Hazardous Chemicals and Pesticides in International Trade: 26–30 May 1997.* June 2.

Earth Negotiations Bulletin. 1997b. *Report of the Fourth Session of the INC for an International Legally Binding Instrument for the Application of the Prior Informed Consent Procedure for Certain Hazardous Chemicals and Pesticides in International Trade: 20–24 October 1997.* October 27.

Earth Negotiations Bulletin. 1998a. *Report of the Fifth Session of the INC for an International Legally Binding Instrument for the Application of the Prior Informed Consent Procedure for Certain Hazardous Chemicals and Pesticides in International Trade: 9–14 March 1998.* March 16.

Earth Negotiations Bulletin. 1998b. *Report of the Conference of Plenipotentiaries on the Convention on the Prior Informed Consent Procedure for Certain Hazardous Chemicals and Pesticides in International Trade: 10–11 September 1998.* September 14, 1998.

Earth Negotiations Bulletin. 1998c. *Report of the First Session of the INC for an International Legally Binding Instrument for Implementing International Action on Certain Persistent Organic Pollutants (POPs): 29 June–3 July 1998.* July 6.

Earth Negotiations Bulletin. 1999a. *Summary of the Fifth Conference of the Parties to the Basel Convention on the Control of Transboundary Movements of Hazardous Wastes and Their Disposal: 6–10 December 1999.* December 13.

Earth Negotiations Bulletin. 1999b. *Report of the Sixth Session of the INC for an International Legally Binding Instrument for the Application of the Prior Informed Consent Procedure for Certain Hazardous Chemicals and Pesticides in International Trade: 12–16 July 1999.* July 19.

Earth Negotiations Bulletin. 1999c. *The Second Session of the Intergovernmental Negotiation Committee for an International Legally Binding Instrument for Implementing International Action on Certain Persistent Organic Pollutants (POPs): 25–29 January 1999.* February 1.

Earth Negotiations Bulletin. 1999d. *Summary of the Third Session of the INC for an International Legally Binding Instrument for Implementing International Action on Certain Persistent Organic Pollutants: 6–11 September 1999.* September 13.

Earth Negotiations Bulletin. 2000a. *Report of the Seventh Session of the INC for an International Legally Binding Instrument for the Application of the Prior Informed Consent Procedure for Certain Hazardous Chemicals and Pesticides in International Trade: 30 October–3 November 2000.* November 6.

Earth Negotiations Bulletin. 2000b. *Summary of the Fourth Session of the Intergovernmental Negotiation Committee for an International Legally Binding In-*

strument for Implementing International Action on Certain Persistent Organic Pollutants: 20–25 March 2000. March 27.

Earth Negotiations Bulletin. 2000c. *Summary of the Fifth Session of the Intergovernmental Negotiation Committee for an International Legally Binding Instrument for Implementing International Action on Certain Persistent Organic Pollutants: 4–9 December 2000.* December 12.

Earth Negotiations Bulletin. 2002a. *Summary of the Sixth Conference of the Parties to the Basel Convention: 9–14 December 2002.* December 16.

Earth Negotiations Bulletin. 2002b. *Summary of the Ninth Session of the Intergovernmental Negotiating Committee for an International Legally Binding Instrument for the Application of the Prior Informed Consent Procedure for Certain Hazardous Chemicals and Pesticides in International Trade: 30 September–4 October 2002.* October 7.

Earth Negotiations Bulletin. 2002c. *Summary of the Sixth Session of the Intergovernmental Negotiation Committee for an International Legally Binding Instrument for Implementing International Action on Certain Persistent Organic Pollutants: 17–21 June 2002.* June 24.

Earth Negotiations Bulletin. 2003. *Summary of the Seventh Session of the Intergovernmental Negotiation Committee for an International Legally Binding Instrument for Implementing International Action on Certain Persistent Organic Pollutants: 14–18 July 2003.* July 21.

Earth Negotiations Bulletin. 2004a. *Summary of the Seventh Conference of the Parties to the Basel Convention: 25–29 October 2004.* November 1.

Earth Negotiations Bulletin. 2004b. *Summary of the Eleventh Session of the INC for an International Legally Binding Instrument for the Application of the Prior Informed Consent Procedure for Certain Hazardous Chemicals and Pesticides in International Trade and the First Conference of the Parties to the Rotterdam Convention: 18–24 September 2004.* September 27.

Earth Negotiations Bulletin. 2005a. *Summary of the Second Conference of the Parties to the Rotterdam Convention on the Prior Informed Consent Procedure for Certain Hazardous Chemicals and Pesticides in International Trade: 27–30 September 2005.* October 3.

Earth Negotiations Bulletin. 2005b. *Summary of the First Conference of the Parties to the Stockholm Convention: 2–6 May 2005.* May 9.

Earth Negotiations Bulletin. 2006a. *Summary of the Eight Conference of the Parties to the Basel Convention: 27 November–1 December 2006.* December 4.

Earth Negotiations Bulletin. 2006b. *Summary of the Third Meeting of the Conference of the Parties to the Rotterdam Convention on the Prior Informed Consent Procedure for Certain Hazardous Chemicals and Pesticides in International Trade: 9–13 October 2006.* October 16.

Earth Negotiations Bulletin. 2006c. *Summary of the Second Conference of the Parties to the Stockholm Convention: 1–5 May 2006.* May 8.

Earth Negotiations Bulletin. 2006d. *Summary of the Second Meeting of the Persistent Organic Pollutant Review Committee of the Stockholm Convention: 6–10 November 2007.* November 13.

Earth Negotiations Bulletin. 2007a. *Summary of the Third Conference of the Parties to the Stockholm Convention: 30 April–4 May 2007.* May 7.

Earth Negotiations Bulletin. 2007b. *Summary of the Third Meeting of the Persistent Organic Pollutant Review Committee of the Stockholm Convention: 19–23 November 2007.* November 26.

Earth Negotiations Bulletin. 2008a. *Summary of the Ninth Conference of the Parties to the Basel Convention: 23–27 June 2008.* June 30.

Earth Negotiations Bulletin. 2008b. *Summary of the Fourth Meeting of the Conference of the Parties to the Rotterdam Convention: 27–31 October 2008.* November 3.

Earth Negotiations Bulletin. 2008c. *Summary of the Fourth Meeting of the Persistent Organic Pollutant Review Committee of the Stockholm Convention: 13–17 October 2008.* October 20.

Earth Negotiations Bulletin. 2009. *Summary of the Fourth Conference of the Parties to the Stockholm Convention: 4–8 May 2009.* May 11.

Eckley, Noelle. 2000. From Regional to Global Assessment: Learning from Persistent Organic Pollutants. ENRP Discussion Paper, Kennedy School of Government, Harvard University.

Eckley, Noelle, and Henrik Selin. 2004. All Talk, Little Action: Precaution and European Chemicals Regulation. *Journal of European Public Policy* 11 (1):78–105.

Emory, Richard W. 2001. Probing the Protections in the Rotterdam Convention on Prior Informed Consent. *Colorado Journal of International Environmental Law and Policy* 12:47–69.

Environment News Service. 2006. World Governments Asked to Pay for Ivory Coast Cleanup. November 24.

Eugster, Martin, Duan Huabo, Li Jinhui, Oshani Perera, Jason Potts, and Wanhua Yang. 2008. *Sustainable Electronics and Electrical Equipment for China and the World: A Commodity Chain Sustainability Analysis of Key Chinese EEE Product Chains.* Winnipeg: International Institute for Sustainable Development.

European Chemical Industry Council. 2006. *Facts and Figures: The European Chemical Industry in a Worldwide Perspective.* Brussels: CEFIC.

European Commission. 2000. *Communication from the Commission on the Precautionary Principle.* Brussels: European Commission.

European Commission. 2001. *White Paper: Strategy for a Future Chemicals Policy.* COM(2001)88 final, February 27. Brussels: European Commission.

European Commission. 2002. *A European Union Strategy for Sustainable Development.* Brussels: European Commission.

Farrell, Alexander E., and Jill Jäger. 2005. *Assessments of Regional and Global Environmental Risks: Designing Processes for the Effective Use of Science in Decisionmaking.* Washington, D.C.: Resources for the Future.

Faure, Michael, and Jürgen Lefevere. 2005. Compliance with Global Environmental Policy. In *The Global Environment: Institutions, Law and Policy*, ed. Regina S. Axelrod, David Leonard Downie, and Norman J. Vig, 125–145. Washington, D.C.: CQ Press.

Fenge, Terry. 2003. POPs and Inuit: Influencing the Global Agenda. In *Northern Lights against POPs: Combatting Toxic Threats in the Arctic*, ed. David Leonard Downie and Terry Fenge, 192–213. Montreal: McGill–Queens University Press.

Forester, William S., and John H. Skinner, eds. 1987. *International Perspectives on Hazardous Waste Management*. London: Academic Press.

Gallup, John Luke, and Jeffrey D. Sachs. The Economic Burden of Malaria. Working paper No. 52, Center for International Development at Harvard University, 2000.

Garcia-Johnson, Ronie. 2000. *Exporting Environmentalism: U.S. Multinational Chemical Corporations in Brazil and Mexico*. Cambridge, Mass.: MIT Press.

Geiser, Kenneth. 2001. *Materials Matter: Towards a Sustainable Materials Policy*. Cambridge, Mass. : MIT Press.

Giesy, J. P., and K. Kannan. 2001. Global Distribution of Perfluorooctane Sulfonate in Wildlife. *Environmental Science and Technology* 35:1339–1342.

Goodman, Peter. 2003. China Serves as Dump Site for Computers: Unsafe Recycling Practice Grows Despite Import Ban. *Washington Post*, February 24, A01.

Greenpeace. 2008. *Toxic Tech: Not in Our Backyard: Uncovering the Hidden Flows of E-waste*. Amsterdam: Greenpeace.

Gupta, Joyeeta. 2008. Global Change: Analyzing Scale and Scaling in Environmental Governance. In *Institutions and Environmental Challenges: Principal Findings, Applications, and Research Frontiers*, ed. Oran R. Young, Leslie A. King, and Heike Schroeder, 225–258. Cambridge, Mass.: MIT Press.

Haas, Ernst B. 1990. *When Knowledge Is Power: Three Models of Change in International Organizations*. Berkeley: University of California Press.

Haas, Peter M. 1990. *Saving the Mediterranean: The Politics of International Environmental Cooperation*. New York: Columbia University Press.

Haas, Peter M. 2004. When Does Power Listen to Truth? A Constructivist Approach to the Policy Process. *Journal of European Public Policy* 11(4):569–592.

Haas, Peter M., Robert O. Keohane, and Marc A. Levy. 1993. *Institutions for the Earth: Sources of Effective International Environmental Protection*. Cambridge, Mass.: MIT Press.

Harding, Sandra. 1991. *Whose Science? Whose Knowledge? Thinking from Women's Lives*. Buckingham: Open University Press.

Harland, David. 1985. Legal Aspects of the Export of Hazardous Products. *Journal of Consumer Policy* 8 (3):209–238.

Harremöes, P., M. MacGarvin, A. Stirling, J. Keys, B. Wynne, and S. G. Vaz, eds. 2002. *The Precautionary Principle in the 20th Century*. London: Earthscan.

Hasenclever, Andreas, Peter Mayer, and Volker Rittberger. 1997. *Theories of International Regimes*. Cambridge: Cambridge University Press.

Holden, A. V. 1970. Monitoring Organochlorine Contamination of the Marine Environment by the Analysis of Residues in Seals. In *Marine Pollution and Sea Life*, ed. M. Ruivo, 266–272. Rome: Food and Agricultural Organization.

Hooghe, Liesbet, ed. 1996. *Cohesion Policy and European Integration: Building Multilevel Governance*. Oxford: Oxford University Press.

Ikonomou, M. G., S. Rayne, and R. F. Addison. 2002. Exponential Increases of the Brominated Flame Retardants, Polybrominated Diphenyl Ethers, in the Canadian Arctic from 1981 to 2000. *Environmental Science and Technology* 36:1886–1892.

Iles, Alastair. 2004. Mapping Environmental Justice in Technology Flows: Computer Waste Impacts in Asia. *Global Environmental Politics* 4 (4):76–107.

Jacobson, J. L., and S. W. Jacobson. 1996. Intellectual Impairment in Children Exposed to Polychlorinated Biphenyls in Utero. *New England Journal of Medicine* 335:783.

Jaffe, Mark. 1988. Tracking the Khian Sea: Port to Port, Deal to Deal. *Philadelphia Inquirer*, July 15, B1.

Jasanoff, Sheila, and Marybeth Long Martello, eds. 2004. *Earthly Politics: Local and Global in Environmental Governance*. Cambridge, Mass.: MIT Press.

Jasanoff, Sheila, and Brian Wynne. 1998. *Science and Decisionmaking*. In *Human Choice and Climate Change*, Vol. 1, *The Societal Framework*, ed. Steve Rayner and Elizabeth L. Malone, 1–87. Columbus, Ohio: Battelle Press.

Jensen, Sören. 1972. The PCB Story. *Ambio* 1 (4):123–131.

Jeyaratnam, J. 1990. Acute Pesticide Poisoning: A Major Global Health Problem. *World Health Statistics Quarterly* 43 (3):139–144.

Joyner, Christoffer C. 1998. *Governing the Frozen Commons: The Antarctic Regime and Environmental Protection*. Columbia: University of South Carolina Press.

Kempel, Willy. 1993. Transboundary Movements in Hazardous Wastes. In *International Environmental Negotiations*, ed. Gunnar Sjöstedt, 48–62. Newbury Park, Calif.: Sage.

Keohane, Robert O. 1989. *International Institutions and State Power: Essays in International Relations Theory*. Boulder, Colo.: Westview Press.

Kinloch, D., and H. Kuhnlein. 1988. Assessment of PCBs in Arctic Foods and Diet: A Pilot Study in Broughton Island, NWT, Canada. *Arctic Medical Research* 47 (Supplement 1):159–162.

Kohler, Pia M. 2006. Science, PIC and POPs: Negotiating the Membership of Chemical Review Committees under the Stockholm and Rotterdam Conventions. *Review of European Community & International Environmental Law* 15 (3):293–303.

Koivurova, Timo, and Leena Heinämäki. 2006. The Participation of Indigenous Peoples in International Norm-making in the Arctic. *Polar Record* 42 (221):101–109.

Koppe, Janna G., and Jane Keys. 2002. PCBs and the Precautionary Principle. In *The Precautionary Principle in the 20th Century*, ed. Poul Harremoës, David Gee, Malcolm MacGarvin, Andy Stirling, Jane Keys, Brian Wynne, and Sophia Guedes Vaz., 64–78, London: Earthscan.

Krasner, Stephen D. 1983. Structural Causes and Regime Consequences: Regimes as Intervening Variables. In *International Regimes*, ed. Stephen D. Krasner, 1–21. Ithaca, N.Y.: Cornell University Press.

Kratochwil, Friedrich V. 1989. *Rules, Norms, and Decisions: On the Conditions of Practical and Legal Reasoning in International Relations and Domestic Affairs.* Cambridge: Cambridge University Press.

Kratochwil, Friedrich. 1993. Contract and Regimes: Do Issue Specificity and Variations of Formality Matter? In *Regime Theory and International Relations*, ed. Volker Rittberger and Peter Mayer, 73–93. Oxford: Clarendon Press.

Krimsky, Sheldon. 2000. *Hormonal Chaos: The Scientific and Social Origins of the Environmental Endocrine Hypothesis.* Baltimore,Md.: Johns Hopkins University Press.

Krueger, Jonathan. 1999. *International Trade and the Basel Convention.* London: Royal Institute for International Affairs.

Krueger, Jonathan, and Henrik Selin. 2002. Governance for Sound Chemicals Management: The Need for a More Comprehensive Global Strategy. *Global Governance* 8 (3):323–342.

Kuhnlein, H. V., and H. M. Chan. 2000. Environment and Contaminants in Traditional Food Systems of Northern Indigenous Peoples. *Annual Review of Nutrition* 20:595–626.

Kuhnlein, Harriet, Laurie H. M. Chan, Grace Egeland, and Oliver Receveur. 2003. Canadian Arctic Indigenous Peoples, Traditional Food System, and POPs. In *Northern Lights against POPs: Combatting Toxic Threats in the Arctic*, ed. David Leonard Downie and Terry Fenge 22–40. Montreal: McGill–Queens University Press,.

Kummer, Katharina. 1995. *International Management of Hazardous Wastes: The Basel Convention and Related Legal Rules.* Oxford: Clarendon Press.

Kummer, Katharina. 1999. Prior Informed Consent for Chemicals in International Trade: The 1998 Rotterdam Convention. *Review of European Community & International Environmental Law* 8 (3):323–330.

Kuratsune, M., H. Yoshimura, Y. Hori, M. Okumura, and Y. Masuda, eds. 1996. *Yusho: A Human Disaster Caused by PCBs and Related Compounds.* Fukuoka: Kyushu University Press.

Larsson, Per, and Lennart Okla. 1989. Atmospheric Transport of Chlorinated Hydrocarbons to Sweden in 1985 Compared to 1973. *Atmospheric Environment* 23 (8):1699–1711.

Layzer, Judith A. 2006. *The Environmental Case: Translating Values into Policy.* 2nd ed. Washington, D.C.: CQ Press.

Lear, Linda. 1997. *Rachel Carson: Witness for Nature.* New York: Holt.

Levy, Marc A., Oran R. Young, and Michael Zürn. 1995. The Study of International Regimes. *European Journal of International Relations* 1 (3):267–331.

Liefferink, Duncan, and Mikael Skou Andersen. 1998. Strategies of the Green Member States in EU Environmental Policy-making. *Journal of European Public Policy* 5 (2):254–270.

Lidskog, Rolf, and Göran Sundqvist. 2002. The Role of Science in Environmental Regimes: The Case of LRTAP. *European Journal of International Relations* 8 (1): 77–101.

Linnér, Björn-Ola. 2003. *The Return of Malthus: Environmentalism and Post-war Population-Resource Crises.* Isle of Harris: The White Horse Press.

Löfstedt, Ragnar E. 2003. Swedish Chemical Regulation: An Overview and Analysis. *Risk Analysis* 23 (2):411–421.

Lönngren, Rune. 1992. *International Approaches to Chemicals Control: A Historical Overview.* Stockholm: National Chemicals Inspectorate/KemI.

Lytle, Mark Hamilton. 2007. *The Gentle Subversive: Rachel Carson, Silent Spring, and the Rise of the Environmental Movement.* Oxford: Oxford University Press.

Mabaso, Wusawenkosi L. H., Brian Sharp, and Christian Lengeler. 2004. Historical Review of Malaria Control in Southern Africa on the Use of Indoor Residual House-Spraying. *Tropical Medicine and International Health* 9 (8):846–856.

Maguire, Steve, and Jaye Ellis. 2005. Redistributing the Burden of Scientific Uncertainty: Implications of the Precautionary Principle for State and Nonstate Actors. *Global Governance* 11 (4):505–526.

Mancini, Francesca, Ariena H. C. Van Bruggen, Janice L. S. Jiggins, Arun C. Ambatipudi, and Helen Murphy. 2005. Acute Pesticide Poisoning among Female and Male Cotton Growers in India. *International Journal of Occupational and Environmental Health* 11 (3):221–232.

Mansbach, Richard W., and John A. Vasquez. 1981. *In Search of Theory: A New Paradigm for Global Politics.* New York: Columbia University Press.

March, James G., and Johan P. Olsen. 1989. *Rediscovering Institutions: The Organizational Basis of Politics.* New York: Free Press.

Marks, Gary. 1992. Structural Policy in the European Community. In *Euro-politics: Institutions and Policymaking in the New European Community,* ed. Alberta M. Sbragia, 191–224. Washington, D.C.: Brookings Institution.

Marks, Gary. 1993. Structural Policy and Multilevel Governance in the EC. In *The State of the European Community,* ed. Alan Cafruny and Glenda Rosenthal, 391–410. New York: Lynne Rienner.

McDorman, Ted L. 2004. The Rotterdam Convention on the Prior Informed Consent Procedure for Certain Hazardous Chemicals and Pesticides in International Trade: Some Legal Notes. *Review of European Community and International Environmental Law* 13 (2):187–200.

Mellanby, Kenneth. 1992. *The DDT Story*. Farnham: British Crop Protection Council.

Millman, Joel. 1988. After Two Years, Ship Dumps Toxic Ash. *New York Times*, November 27, A22.

Mingst, Karen A. 1981. The Functionalist and Regime Perspectives: The Case of Rhine River Cooperation. *Journal of Common Market Studies* 20 (2):161–173.

Mitchell, Ronald B, William C. Clark, David W. Cash, and Nancy M. Dickson, eds. 2006. *Global Environmental Assessments: Information and Influence*. Cambridge, Mass.: MIT Press.

Muir, Derek C. G., Ross J. Norstrom, and Mary Simon. 1988. Organochlorine Contaminants in Arctic Marine Food Chains: Accumulation of Specific Polychlorinated Biphenyls and Chlordane-Related Compounds. *Environmental Science and Technology* 22 (9):1071–1079.

Muir, Derek C. G., et al. 1990. Geographic Variation of Chlorinated Hydrocarbons in Burbot (Lota lota) from Remote Lakes and Rivers in Canada. *Archives of Environmental Contamination and Toxicology* 19 (4):530–542.

Najam, Adil. 2005. The View from the South: Developing Countries in Global Environmental Politics. In *The Global Environment: Institutions, Law and Policy*, ed. Regina S. Axelrod, David Leonard Downie, and Norman J. Vig, 225–243. Washington, D.C.: CQ Press.

Nepal. 2007. *National Implementation Plan for the Stockholm Convention on Persistent Organic Pollutants (POPs)*.

Nwilene, F. E., K. F. Nwanze, and A. Youdeowei. 2008. Impact of Integrated Pest Management on Food and Horticultural Crops in Africa. *Entomologia Experimentalis et Applicata* 128 (3):355–363.

Oberthür, Sebastian, and Thomas Gehring, eds. 2006. *Institutional Interaction in Global Environmental Governance: Synergy and Conflict among International and EU Policies*. Cambridge, Mass.: MIT Press.

Oberthür, Sebastian, and Thomas Gehring. 2008. Interplay: Exploring Institutional Interaction. In *Institutions and Environmental Challenges: Principal Findings, Applications, and Research Frontiers*, ed. Oran R. Young, Leslie A. King, and Heike Schroeder, 187–223. Cambridge, Mass.: MIT Press.

Oehme, M., and S. Manø. 1984. The Long-Range Transport of Organic Pollutants to the Arctic. *Fresenius' Zeitschrift für Analytische Chemie* 319:141–146.

O'Neill, Kate. 2000. *Waste Trading among Rich Nations: Building a New Theory of Environmental Regulation*. Cambridge, Mass.: MIT Press.

Organisation for Economic Cooperation and Development. 1973a. *Decision of the Council on Protection of the Environment by Control of Polychlorinated Biphenyls*. February 13. Paris: OECD.

Organisation for Economic Cooperation and Development. 1973b. *Polychlorinated Biphenyls: Their Use and Control*. Paris: OECD Environment Directorate.

Organisation for Economic Cooperation and Development. 1981. *OECD and Chemicals Control: The High Level Meeting of the Chemicals Group of the Environment Committee, 1980*. Paris: OECD.

Organisation for Economic Cooperation and Development. 1985. *Transfrontier Movements of Hazardous Wastes*. Paris: OECD.

Organisation for Economic Cooperation and Development. 2001. *OECD Environmental Outlook for the Chemicals Industry*. Paris: OECD.

O'Riordan, Timothy, and James Cameron, eds. 1994. *Interpreting the Precautionary Principle*. London: Earthscan.

Paarlberg, Robert L. 1993. Managing Pesticide Use in Developing Countries. In *Institutions for the Earth: Sources of Effective International Environmental Protection*, ed. Peter M. Haas, Robert O. Keohane, and Marc A. Levy, 309–350. Cambridge, Mass.: MIT Press.

Pallemaerts, M. 1988. Developments in International Pesticide Regulation. *Environmental Policy and Law* 18 (3):62–69.

Pallemaerts, Marc. 2003. *Toxics and Transnational Law: International and European Regulation of Toxic Substances as Legal Symbolism*. Portland, Ore.: Hart Publishing.

Palmer, Michael G. 2007. The Case of Agent Orange. *Contemporary Southeast Asia* 29 (1):172–192.

Parson, Edward A. 2003. *Protecting the Ozone Layer: Science and Strategy*. Oxford: Oxford University Press.

Pellow, David Naguib. 2007. *Resisting Global Toxics: Transnational Movements for Environmental Justice*. Cambridge, Mass.: MIT Press.

Princen, Thomas, and Matthias Finger. 1995. *Environmental NGOs in World Politics*. London: Routledge.

Raffensperger, C., and J. Tickner, eds. 1999. *Protecting Public Health and the Environment: Implementing the Precautionary Principle*. Washington, D.C.: Island Press.

Raustiala, Kal, and David G. Victor. 2004. The Regime Complex for Plant Genetic Resources. *International Organization* 58 (2):277–309.

Reiersen, Lars-Otto, Simon Wilson, and Vitaly Kimstach. 2003. Circumpolar Perspectives on Persistent Organic Pollutants: The Arctic Monitoring and Assessment Programme. In *Northern Lights against POPs: Combatting Toxic Threats in the Arctic*, ed. David Leonard Downie and Terry Fenge, 60–86. Montreal: McGill–Queens University Press.

Renckens, Stefan. 2008. Yes, We Will! Voluntarism in US E-Waste Governance. *Review of European Community and International Environmental Law* 17 (3): 286–299.

Report of a New Chemical Hazard. 1966. *New Scientist* 32 (525):612.

Ringius, Lasse. 2001. *Radioactive Disposal at Sea: Public Ideas, Transnational Policy Entrepreneurs, and Environmental Regimes.* Cambridge, Mass.: MIT Press.

Roberts, D. R., S. Manguin, and J. Mouchet. 2000. DDT House Spraying and Re-emerging Malaria. *Lancet* 356 (9226):330–332.

Rodan, Bruce, et al. 1999. Screening for Persistent Organic Pollutants: Techniques to Provide a Scientific Basis for POPs Criteria in International Negotiations. *Environmental Science and Technology* 33 (20):3482–3488.

Rogan, Walter J., and Aimin Chen. 2005. Health Risks and Benefits of Bis(4-Cholorophenyl)-1,1,1-Trichloroethane (DDT). *Lancet* 366 (9487):763–773.

Rosendal, G. Kristin. 2001. Impact of Overlapping International Regimes: The Case of Biodiversity. *Global Governance* 7 (2):95–117.

Rothstein, Henry F. 2003. Neglected Risk Regulation: The Institutional Attenuation Phenomenon. *Health Risk and Society* 5 (1):85–103.

Sachs, Jeffrey D. 2002. A New Global Effort to Control Malaria. *Science* 298 (5591):122–124.

Sandin, P., M. Peterson, S. O. Hansson, C. Rudén, and A. Juthe. 2002. Five Charges against the Precautionary Principle. *Journal of Risk Research* 5 (4):287–299.

Scheringer, Martin. 2007. Towards an Intergovernmental Panel on Chemical Pollution (IPCP). *Chemosphere* 67 (9):1682–1683.

Schörling, Inger. 2003. The Greens Perspective on EU Chemicals Regulation and the White Paper. *Risk Analysis* 3 (2):405–409.

Schörling, Inger, and Gunnar Lund. 2004. *The Only Planet Guide to the Secrets of Chemicals Policy in the EU: REACH-What Happened and Why?* Brussels: Greens/European Free Alliance in the European Parliament.

Schroeder, Heike, Leslie A. King, and Simon Tay. 2008. Contributing to the Science-Policy Interface: Policy Relevance of Findings on the Institutional Dimensions of Global Environmental Change. In *Institutions and Environmental Challenges: Principal Findings, Applications, and Research Frontiers*, ed. Oran R. Young, Leslie A. King, and Heike Schroeder, 261–275. Cambridge, Mass.: MIT Press.

Schulberg, Francine. 1979. United States Export of Products Banned for Domestic Use. *Harvard International Law Journal* 20 (2):331–383.

Selin, Henrik. 2000. *Towards International Chemical Safety: Taking Action on Persistent Organic Pollutants (POPs).* Linköping: Linköping Studies in Arts and Science, no. 211.

Selin, Henrik. 2003. Regional POPs Policy: The UNECE/CLRTAP POPs Agreement. In *Northern Lights against POPs: Combatting Toxic Threats in the Arctic*, ed. David Leonard Downie and Terry Fenge, 111–132. Montreal: McGill–Queen's University Press.

Selin, Henrik. 2007. Coalition Politics and Chemicals Management in a Regulatory Ambitious Europe. *Global Environmental Politics* 7 (3):63–93.

Selin, Henrik. 2009a. Transatlantic Politics of Chemicals Management. In *Enlarging Transatlantic Environment and Energy Politics: Comparative and International Perspectives*, ed. Miranda A. Schreurs, Henrick Selin, and Stacey D. VanDeveer, 57–74. Aldershot: Ashgate.

Selin, Henrik. 2009b. *Managing Hazardous Chemicals: Long-Range Challenges.* Boston University: Frederick S. Pardee Center for the Study of the Longer-Range Future.

Selin, Henrik, and Noelle Eckley. 2003. Science, Politics, and Persistent Organic Pollutants: Scientific Assessments and Their Role in International Environmental Negotiations. *International Environmental Agreements: Politics, Law and Economics* 3 (1):17–42.

Selin, Henrik, and Olof Hjelm. 1999. The Role of Environmental Science and Politics in Identifying Persistent Organic Pollutants for International Regulatory Actions. *Environmental Reviews* 7 (2):61–68.

Selin, Henrik, and Noelle Eckley Selin. 2008. Indigenous Peoples in International Environmental Cooperation: Arctic Management of Hazardous Substances. *Review of European Community and International Environmental Law* 17 (1):72–83.

Selin, Henrik, and Stacy D. VanDeveer. 2003. Mapping Institutional Linkages in European Air Pollution Politics. *Global Environmental Politics* 3 (3):14–46.

Selin, Henrik, and Stacy D. VanDeveer. 2004. Baltic Sea Hazardous Substances Management: Results and Challenges. *Ambio* 33 (3):153–160.

Selin, Henrik, and Stacy D. VanDeveer. 2006. Raising Global Standards: Hazardous Substances and E-Waste Management in the European Union. *Environment* 48(10): 6–18.

Selin, Henrik, and Stacy D. VanDeveer, eds. 2009. *Changing Climates in North American Politics: Institutions, Policymaking and Multilevel Governance.* Cambridge, Mass.: MIT Press.

Selin, Noelle Eckley. 2006. From Regional to Global Information: Assessment of Persistent Organic Pollutants. In *Global Environmental Assessments: Information and Influence*, ed. Ronald B. Mitchell, William C. Clark, David W. Cash, and Nancy M. Dickson, 175–199. Cambridge, Mass.: MIT Press.

Selin, Noelle Eckley, and Henrik Selin. 2006. Global Politics of Mercury Pollution: The Need for Multi-Scale Governance. *Review of European Community and International Environmental Law* 15 (3):258–269.

Shaw, Malcolm. 1983. The United Nations Convention on Prohibitions or Restrictions on the Use of Certain Conventional Weapons, 1981. *Review of International Studies* 9 (1):109–121.

Shearer, Russel, and Siu-Ling Han. 2003. Canadian Research and POPs: The Northern Contaminants Program. In *Northern Lights against POPs: Combatting Toxic Threats in the Arctic*, ed. David Leonard Downie and Terry Fenge, 41–59. Montreal: McGill–Queens University Press.

Shifrin, Neil S., and Amy P. Toole. 1998. Historical Perspective on PCBs. *Environmental Engineering Science* 15 (3):247–257.

Skodvin, Tora, and Steinar Andresen. 2006. Leadership Revisited. *Global Environmental Politics* 6 (3):13–27.

Sonak, Sangeeta, Mahesh Sonak, and Asha Giriyan. 2008. Shipping Hazardous Waste: Implications for Economically Developing Countries. *International Environmental Agreements: Politics, Law and Economics* 8 (2):143–159.

Srinivas Rao, C. H., V. Venkateswarlu, T. Surender, Michael Eddleston and Nich A. Buckley. 2005. Pesticide Poisoning in South India: Opportunities for Prevention and Improved Medical Treatment. *Tropical Medicine and International Health* 10 (6):581–588.

Stellman, Jeanne Mager, Steve D. Stellman, Richard Christian, Tracy Weber, and Carrie Tomasallo. 2003. The Extent and Patterns of Usage of Agent Orange and Other Herbicides in Vietnam. *Nature* 422 (6933):681–687.

Stockholm Convention Secretariat. 2006. *Study on Improving Cooperation and Synergies between the Secretariats of the Basel, Rotterdam and Stockholm Conventions.* Geneva: Stockholm Convention Secretariat.

Stockholm Convention Secretariat. 2009a. *Report of the Export Group on Assessment of the Production and Use of DDT and its Alternatives for Disease Vector Control.* Geneva: Stockholm Convention Secretariat.

Stockholm Convention Secretariat. 2009b. *Global Status of DDT and Its Alternatives for Use in Vector Control to Prevent Disease.* Geneva: Stockholm Convention Secretariat.

Stokke, Olav Schram. 2001. *Governing High Seas Fisheries: The Interplay of Global and Regional Regimes.* Oxford: Oxford University Press.

Susskind, Lawrence E. 1994. *Environmental Diplomacy. Negotiating More Effective Global Agreements.* Oxford: Oxford University Press.

Sweden. 1996. Short Chain Chlorinated Paraffins. Paper presented at the ad hoc Preparatory Working Group on Persistent Organic Pollutants Meeting, October 21–23, Aylmer, Canada.

Sweden. 1997a. Risk Assessment of Lindane. Paper presented at the Working Group on Strategies Meeting, January 20–24, Geneva, Switzerland.

Sweden. 1997b. Short Chain Chlorinated Paraffins—Additional Risk Information. Paper presented at the Working Group on Strategies Meeting, January 20–24, Geneva, Switzerland.

Swedish Chemicals Inspectorate. 2006. *National Implementation Plan for the Stockholm Convention on Persistent Organic Pollutants for Sweden.* May. Sundbyberg, Sweden: Swedish Chemicals Inspectorate.

Swedish Environmental Protection Agency. 1990. *Persistent Organic Compounds in the Marine Environment, Report 3690.* Stockholm: Swedish Environmental Protection Agency.

Tanzania. 2005. *National Implementation Plan (NIP) for the Stockholm Convention on Persistent Organic Pollutants (POPs).*

Tolba, Mostafa K. 1990. The Global Agenda and the Hazardous Wastes Challenge. *Marine Policy* 14 (3):205–209.

Tremblay, Jean-François. 2007. China's Cancer Villages. *Chemical and Engineering News* 85 (44):18–21.

Tupper, Karl A. 2009. Endosulfan Faces Growing Bans around the World. *PAN Magazine*, Spring.

Underdal, Arild. 1994. Leadership Theory: Rediscovering the Arts of Management. In *International Multilateral Negotiation: Approaches to the Management of Complexity*, ed. I. W. Zartman, 178–197. San Francisco: Jossey-Bass.

Underdal, Arild. 2008. Determining the Causal Significance of Institutions: Accomplishments and Challenges. In *Institutions and Environmental Challenges: Principal Findings, Applications, and Research Frontiers*, ed. Oran R. Young, Leslie A. King, and Heike Schroeder, 49–78. Cambridge, Mass.: MIT Press.

United Nations Economic Commission for Europe. 1989. *Report of the Eighth Session*. EB.AIR/WG.1/12, September 18.

United Nations Economic Commission for Europe. 1990. *Report of the Eighth Session of the Executive Body.* ECE/EB. AIR 24 (December):11.

United Nations Economic Commission for Europe. 1994. *Draft Executive Summary of the State of Knowledge Report of the Task Force on Persistent Organic Pollutants led by Canada and Sweden.* EB.AIR/WG.6/R.20/Add.1, April 25.

United Nations Economic Commission for Europe. 1999. *Draft 2000 Work-Plan for the Implementation of the Convention on Long-Range Transboundary Air Pollution.* EB.AIR/1999/7/Add.1, October 12.

United Nations Economic Commission for Europe. 2000. *Report of the Thirty-Second Session.* EB.AIR/WG.5/66, September 12.

United Nations Economic Commission for Europe. 2004. *Report of the Twenty-First Session of the Executive Body.* ECE/EB.AIR/79/Add.2, January 23.

United Nations Economic Commission for Europe. 2007a. *Report of the Executive Body on its Twenty-Fourth Session, held in Geneva from 11–14 December 2006.* ECE/EB. AIR 89 (March):1.

United Nations Economic Commission for Europe. 2007b. *Negotiation of a Revised of New Protocol on Persistent Organic Pollutants.* ECE/EB.AIR/WG.5/2007/14, July.

United Nations Environment Programme. 2002a. *Atlas of International Freshwater Agreements.* Kenya: United Nations Environment Programme.

United Nations Environment Programme. 2002b. *Global Trends in Generation and Transboundary Movements of Hazardous Wastes and Other Wastes.* Basel Convention Series/SBC No. 02/14.

United Nations Environment Programme. 2002c. *Basel Convention to Adopt Strategic Plan for Global Action on Hazardous and Other Wastes. UNEP Information Note* 2002 (33):9.

United Nations Environment Programme. 2002d. *Partnership with Industry.* UNEP/CHW.6/32/Add.1, October 31.

United Nations Environment Programme. 2005. E-waste, the Hidden Side of IT Equipment's Manufacturing and Use. *Environment Alert Bulletin*, January.

United Nations Environment Programme Chemicals. 2004. *Overview and Summary of Outcomes from the GEF-MSP Sub-Regional Workshops on Support for the Implementation of the Stockholm Convention on POPs Working Groups.* Geneva: UNEP Chemicals.

United Nations Environment Programme Governing Council. 1995. *Decision 18/32: Persistent Organic Pollutants*, May 25.

United Kingdom. 1996. *Review of Risk Characterisation Information on Selected Persistent Organic Pollutants.* Paper presented at the CLRTAP ad hoc Preparatory Working Group Persistent Organic Pollutants Meeting, October 21–23, Aylmer, Canada.

United States. 1997a. Environmental Fate and Transport of Pentachlorophenol, Review of Screening Criteria Data for Persistent Organic Pollutants. Paper presented at the Working Group on Strategies Meeting, October 20–24, Geneva, Switzerland.

United States. 1997b. What Does the Science Show Regarding Pentachlorophenol as a UNECE LRTAP POP? Paper presented at the Working Group on Strategies Meeting, October 20–24, Geneva, Switzerland.

Victor, David G. 1998. Learning by Doing in the Nonbinding International Regime to Manage Trade in Hazardous Chemicals and Pesticides. In *The Implementation and Effectiveness of International Environmental Commitments: Theory and Practice*, ed. David G. Victor, Kal Raustiala, and Eugene B. Skolnikoff, 221–281. Cambridge, Mass.: MIT Press.

Victor, David G., Kal Raustiala, and Eugene B. Skolnikoff, eds. 1998. *The Implementation and Effectiveness of International Environmental Commitments: Theory and Practice.* Cambridge, Mass.: MIT Press.

Vietnam. 2006. Vietnam National Implementation Plan for Stockholm Convention on Persistent Organic Pollutants.

Wania, Frank, and Donald Mackay. 1993. Global Fractionation and Cold Condensation of Low Volatility Organochlorine Compounds in Polar Regions. *Ambio* 22 (1):10–18.

Wania, Frank, and Donald Mackay. 1996. Tracking the Distribution of Persistent Organic Pollutants: Control Strategies for These Contaminants Will Require a Better Understanding of How They Move around the Globe. *Environmental Science and Technology* 30 (9):390A–396A.

Watt-Cloutier, Sheila. 2003. The Inuit Journey towards a POPs-Free World. In *Northern Lights against POPs: Combatting Toxic Threats in the Arctic*, ed. David Leonard Downie and Terry Fenge, 256–267. Montreal: McGill–Queen's University Press.

Wettestad, Jørgen. 2001. Designing Effective Environmental Regimes: The Conditional Keys. *Global Governance* 7 (3):317–341.

Wettestad, Jørgen. 2002. *Clearing the Air: European Advances in Tackling Acid Rain and Atmospheric Pollution.* Aldershot: Ashgate.

World Health Organization. 2000. *WHO Expert Committee on Malaria: Twentieth Report.* Geneva: WHO.

World Health Organization. 2006. *Pesticides and Their Application: For the Control of Vectors and Pests of Public Health Importance.* 6th ed., Geneva: WHO.

World Summit on Sustainable Development. 2002. *Plan of Implementation of the World Summit on Sustainable Development.* Johannesburg: WSSD.

Yamey, Gavin. 2001. Global Campaign to Eradicate Malaria. *BMJ (Clinical Research Ed.)* 322 (7296):1191–1192.

Yang, Wanhua. 2008. Regulating Electrical and Electronic Wastes in China. *Review of European Community and International Environmental Law* 17 (3):337–346.

Young, Oran R. 1989. *International Cooperation: Building Regimes for Natural Resources and the Environment.* Ithaca, N.Y.: Cornell University Press.

Young, Oran R. 1991a. *International Cooperation: Building Regimes for Natural Resources and the Environment.* 2nd ed. Ithaca, N.Y.: Cornell University Press.

Young, Oran R. 1991b. Political Leadership and Regime Formation: On the development of Institutions in International Society. *International Organization* 45 (3):281–308.

Young, Oran R. 1996. Institutional Linkages in International Society: Polar Perspectives. *Global Governance* 2 (1):1–24.

Young, Oran R. 2002. *The Institutional Dimensions of Environmental Change: Fit, Interplay, and Scale.* Cambridge, Mass.: MIT Press.

Young, Oran R. 2008a. Institutions and Environmental Change: The Scientific Legacy of a Decade of IDGEC Research. In *Institutions and Environmental Challenges: Principal Findings, Applications, and Research Frontiers,* ed. Oran R. Young, Leslie A. King, and Heike Schroeder, 3–45. Cambridge, Mass.: MIT Press.

Young, Oran R. 2008b. The Architecture of Global Environmental Governance: Bringing Science to Bear on Policy. *Global Environmental Politics* 8 (1):14–32.

Young, Oran R., Leslie A. King, and Heike Schroeder, eds. 2008. *Institutions and Environmental Challenges: Principal Findings, Applications, and Research Frontiers.* Cambridge, Mass.: MIT Press.

Whiteside, Kerry H. 2006. *Precautionary Politics: Principle and Practice in Confronting Environmental Risk.* Cambridge, Mass.: MIT Press.

Wolf, Amanda. 2000. Informed Consent: A Negotiated Formula for Trade in Risky Organisms and Chemicals. *International Negotiation* 5 (3):485–521.

Index

Italicized "t" refers to table, italicized "b" refers to box.